Encyclopedia of
Mathematics and Society

Space, Science, and the Environment

Encyclopedia of Mathematics and Society

Space, Science, and the Environment

Editors
Sarah J. Greenwald
and
Jill E. Thomley
Applachian State University

SALEM PRESS
A Division of EBSCO Publishing
Ipswich, Massachusetts

Cover Photo: © Mark Garlick/Science Photo Library/Corbis

Copyright © 2013, by Salem Press, A Division of EBSCO Publishing, Inc.
All rights reserved. No part of this work may be used or reproduced in any manner whatsoever or transmitted in any form or by any means, electronic or mechanical, including photocopy, recording, or any information storage and retrieval system, without written permission from the copyright owner. For permissions requests, contact proprietarypublishing@ebscohost.com.

ISBN 978-1-4298-3754-5

Printed in the United States of America

Contents

List of Contributors	ix
Accident Reconstruction	1
Animals	2
Astronomy	8
Bees	11
Black Holes	13
Brain	15
Carbon Dating	20
Carbon Footprint	22
Caves and Caverns	25
Chemotherapy	27
Climate Change	28
Clouds	34
Coral Reefs	35
Crystallography	37
Deforestation	38
Diagnostic Testing	41
Disease Survival Rates	43
Doppler Radar	44
Earthquakes	46
Elementary Particles	47
Elevation	49
Energy	51
Extinction	53

Farming	55
Fertility	58
Fingerprints	60
Firearms	61
Floods	64
Forest Fires	66
Game Theory	68
Genetics	70
Geometry of the Universe	73
Geothermal Energy	78
Gravity	80
Green Mathematics	82
Growth Charts	84
Hurricanes and Tornadoes	85
Infectious Diseases, Tracking	88
Intelligence Quotients	90
Interplanetary Travel	92
Joints	95
LD50/Median Lethal Dose	96
Life Expectancy	98
Light	99
Lightning	101
Maps	103
Molecular Structure	106
Moon	107
Nanotechnology	110
Nervous System	112
Nutrition	113
Pacemakers	116
Planetary Orbits	117
Plate Tectonics	118
Predator–Prey Models	120
Pregnancy	122
Psychological Testing	124
Radiation	126
Recycling	127

Relativity	129
Satellites	132
Solar Panels	133
Stethoscopes	135
Sunspots	136
Synchrony and Spontaneous Order	137
Telescopes	139
Temperature	142
Tides and Waves	144
Transplantation	146
Ultrasound	147
Universal Constants	148
Viruses	150
Vision Correction	151
Volcanoes	153
Water Quality	154
Weather Forecasting	156
Weather Scales	158
Weightless Flight	160
Wind and Wind Power	162
Resource Guide	**165**
Index	**169**

List of Contributors

John G. Alford
  Sam Houston State University
Or Syd Amit
  Boston College
Matt Boelkins
  Grand Valley State University
Sarah Boslaugh
  Washington University School of Medicine
Marek Brabec
  Academy of Sciences of the Czech Republic
David Brink
  University College Dublin, Ireland
Jason L. Churchill
  Cleo Research Associates
Kumer Pial Das
  Lamar University
Maria Droujkova
  Natural Math
Steven R. Edwards
  Southern Polytechnic State University
Daniel J. Galiffa
  Penn State Erie, The Behrend College
Angela Gallegos
  Occidental College
Joaquim Alves Gaspar
  Universidade de Lisboa
Sommer Gentry
  United States Naval Academy
Darren Glass
  Gettysburg College
Sarah J. Greenwald
  Appalachian State University
Alexander A. Gurshtein
  Mesa State College

Juan B. Gutierrez
  University of Miami
Simone Gyorfi
  O. Goga High School, Jibou, Romania
Gareth Hagger-Johnson
  The University of Leeds
Ziaul Hasan
  University of Illinois, Chicago
Holly Hirst
  Appalachian State University
David I. Kennedy
  Shippensburg University of Pennsylvania
Christine Klein
  Independent Scholar
Michael Klucznik
  St. Bonaventure University
Rick Kreminski
  Colorado State University, Pueblo
Bill Kte'pi
  Independent Scholar
Maria Elizete Kunkel
  University of Ulm, Germany
James Landau
  Independent Scholar
Silvia Liverani
  University of Bristol
Chad T. Lower
  Pennsylvania College of Technology
Margaret MacDougall
  University of Edinburgh Medical School
Ashwin Mudigonda
  Universal Robotics Inc.
Andrew Nevai
  University of Central Florida

Eoin O'Connell
  Deakin University
Julian Palmore
  University of Illinois at Urbana-Champaign
Thomas J. Pfaff
  Ithaca College
Mark Roddy
  Seattle University
Maria Elizabeth S. Rodrigues
  University of Ulm, Germany
David C. Royster
  University of Kentucky
Douglas Rugh
  Independent Scholar
Dorry Segev
  Johns Hopkins University School of Medicine
Barbara A. Shipman
  University of Texas at Arlington
Kevin L. Shirley
  Appalachian State University

Ravi Sreenivasan
  University of Mysore
Catherine Stenson
  Juniata College
Jill E. Thomley
  Appalachian State University
Todd Timmons
  University of Arkansas, Fort Smith
Marcella Bush Trevino
  Independent Scholar
Daniela Velichova
  Slovak Technical University in Bratislava
Karen Doyle Walton
  DeSales University
Bethany White
  University of Western Ontario
Qiang Zhao
  Texas State University

Accident Reconstruction

Category: Travel and Transportation.
Fields of Study: Algebra; Data Analysis and Probability; Measurement.
Summary: Accidents can be mathematically reconstructed to model accident risk and to improve safety equipment designs.

Accident reconstruction is important for understanding how accidents happen and for preventing accidents in the future. Principles and techniques from physics, mathematics, engineering, and other sciences are used to quantify critical variables and calculate others. For example, the initial speed of a suddenly braking vehicle can be determined by mathematically analyzing tire skid and yaw marks. The length of skid marks is a function of vehicle velocity and the amount of friction between the wheels and the road surface. In the case of yaw or circular motion, the radius of the yaw mark is also a factor in the calculation, as well as the elevation of the road. Speed can also be calculated from the trajectories, angles, and other characteristics of objects struck by a speeding vehicle, or between two or more colliding vehicles. Investigators may use distances and angles to determine the original positions of passengers who have been ejected from a vehicle. For more complex modeling, mathematicians, engineers, and other accident reconstructors rely on principles and equations from physics, such as those governing energy and momentum, as well as vehicle specifications, mechanical failure analyses, geometric characteristics of highways, and quantification of visibility, perception, and reaction. Data from both real accidents and staged collisions, along with statistically designed safety analyses and other methods such as stochastic modeling, are often used to construct accident simulations and visualizations for use in a wide variety of contexts, including legal proceedings. Actuaries use accident data to model accident risk, which in turn influences insurance rates and public policy, such as seat belt and helmet laws.

Modeling Accident Reconstructions

Accidents related to travel and transportation can have a variety of negative consequences including personal injury and death. The analysis of accidents can lead to improved designs of vehicles and reduced fatalities as well as warning travelers about potential risks of travel. In reconstructing accidents, evidence from photographs, videos, eyewitnesses, or police reports is collected. Decision trees are used to ask questions at each stage of reconstruction and help decide the closest accident scenario dictated by the available evidence. In such reconstructions, probability must be assigned for the likely cause of the accident and for the particular accident type among the possible accident scenarios based on the available evidence. Stochastic modeling is used to help solve such problems in accident reconstructions.

Uses of Accident Reconstructions

Another important aspect of accident reconstructions is to estimate the probability of occurrence of various types of injuries one may suffer in accidents. Such probability estimates are used to help calculate travel insurance. By nature, accidents happen randomly and—since the types of injuries suffered in accidents also vary randomly—it is important to model accident types and predict the kinds of injuries one may suffer in different accident types. Such models can help prepare communities with the optimal number of emergency services and also help doctors prepare for any unique types of injuries they are likely to deal with.

A typical problem is determining the types of special medical facilities that should be established to deal with travel-related accidents in a city. Such problems require stochastic modeling based on past data, which will help in simulating different types of accidents. Simulations help in planning emergency services to deal with accidents. Accident reconstructions may also help in forecasting the number of accidents of different types likely to happen in the near future, which may lead to better planning of the health, emergency, and disaster management facilities in the city.

Safety and Design Using Accident Reconstructions

Accident reconstructions also may help in improving vehicle design. Incorporating safety devices in vehicles is also a very important aspect of design. Safety devices, which help in avoiding severe injuries to passengers because of accidents, are designed with the help of accident reconstruction and are always a matter of high priority. Simulations can be used to develop sensors that can give an early warning about impending accidents or reduce the speeds of vehicles—thereby reducing the severity of an accident. In creating such designs, mathematical optimization methods are used to determine the optimal cost and space to be allotted. Another crucial application of accident reconstruction and accident modeling is driver training. Sophisticated simulators can be used to simulate different accident scenarios and train drivers to react appropriately to each situation in real time. These simulators are based on algorithms and use random number generators to simulate accident situations. Well-developed algorithms that closely simulate real accidents are needed to reduce—or even eliminate—major accidents.

Further Reading

Brach, Raymond, and R. Matthew Branch. *Vehicle Accident Analysis and Reconstruction Methods*. Warrendale, PA: SAE International, 2005.

Franck, Harold, and Darren Franck. *Mathematical Methods for Accident Reconstruction: A Forensic Engineering Perspective*. Boca Raton, FL: CRC Press, 2009.

Ravi Sreenivasan

Animals

Category: Weather, Nature, and Environment.
Fields of Study: Algebra; Communication; Geometry; Measurement.
Summary: Principles of engineering, physics, and mathematics are demonstrated by the physiology, movement, and behavior of animals.

Animals, including human beings, are living organisms that belong to the domain Eukaryota (having complex cellular structures enclosed with membranes) and the kingdom Animalia. Within this taxonomy, the kingdom is defined by several characteristics, including internal digestion of food (called "heterotrophism") and the ability to move using its own energy in at least some stages of life (called "motility"). Some say that what distinguishes humans from other animals is mathematical ability. However, researchers have studied a diverse range of mathematical concepts as they relate to animal behavior and have found evidence of abilities such as symbolic calculation, efficiency in locomotion, and synchrony. There are questions about whether these findings are biased perceptions of mathematical significance. Many mathematical patterns and symmetry can also be found in the structure of animals, ranging from their cellular tissue to their coat patterns. Some of the motivation behind the development of many statistical measures and methods, such as standard deviation and regression, was to characterize natural variability and associations in animal species.

Biological Systematics, Set Theory, and Logic

Biological systematics is the field that describes and names living organisms, provides their classifications and keys for identification, and situates classes of organisms within evolutionary history and modern adaptations. In particular, classification of organisms (called "taxonomy") is an empirical science, where description of classes is the final step in the discovery and description of organisms. Existing biological classifications may include the ranks of domain, kingdom, phylum, class, order, family, genus, and species.

The definition of the kingdom Animalia is intensional definition—it specifies necessary and sufficient conditions for belonging to the set of animals. The particular subclass of definitions used in systematics to define animals is called definition by genus and differentia. Such definitions rely on a structure of sets, subsets, and supersets as well as their differentiating conditions. For example, defining negative numbers as the set of rational numbers that are less than zero, mathematicians use the superset of rational numbers (defined elsewhere) and the differentiating condition of being less than zero. Animalia is one of several kingdoms (subsets) of the domain Eukaryota, differentiated from other kingdoms by particular conditions.

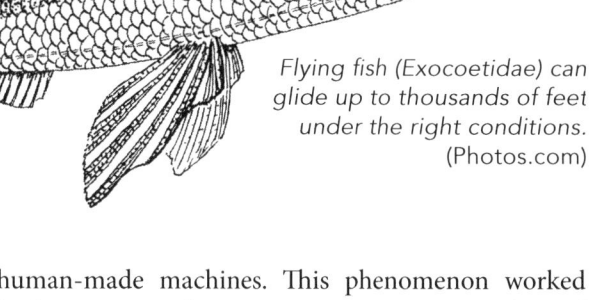

Flying fish (Exocoetidae) can glide up to thousands of feet under the right conditions. (Photos.com)

Careful decisions are made in the organization of kingdoms and in defining differentiated conditions. For example, if only the conditions of internal digestion of food and motility were used, the Venus flytrap would be considered an animal rather than a plant. However, plants are also differentiated by the sufficient condition of having plastids, such as chloroplast, in their cells. Internal digestion of food and motility are necessary but not sufficient conditions for declaring an organism an animal. There are historical and modern systems defining anywhere from two to eight kingdoms of living organisms, depending on the necessary and sufficient conditions used for definitions.

Animal Tissue Structures

All animal cells have extracellular matrix, the boundary that can serve many functions, including exchanging substances between cells, segregating tissues, and anchoring cells. Animal cells typically form tissues; groups of cells carry out particular functions within animal bodies. There are four types of animal tissues, defined by their functions: muscle, nervous, epithelial, and connective. Cells within tissues and tissues within organs may be tessellated (filling space or surface infinitely, without gaps).

Tissue engineering is an interdisciplinary field combining biology, material science, chemistry, and engineering to re-create, change, or replace tissues. It pays special attention to the mechanical and structural properties of tissues, often modeled mathematically before being implemented in the lab.

Technological Metaphors and Models

Beginning in the Renaissance, it was common for people to conceptualize living organisms in terms of human-made machines. This phenomenon worked both ways, since human constructions were informed by new understandings and observations of nature. During the Renaissance, animal tissues and organs were seen as combinations of relatively simple mechanisms such as levers. Attempts were made to imitate some functions of animals in construction, such as making bird-like wings. This analytic approach informed the development of scientific methods in biology—in contrast with a holistic view of living things as having a completely different nature from human-made mechanisms. In the seventeenth century, this philosophical approach of modeling animals on machines was supported by such influential scientists as Galileo Galilei, René Descartes, and Isaac Newton.

Engineering and mathematics developed along with explanations in biology. Developments in steam technology introduced the ideas of energy and work, which, in turn, led to the analysis of gas and liquid pressures as explanations of the interaction of tissues and organs in animal bodies. The metaphors of heart or cellular structures as pumps—or kidneys and the liver as filters—persist to this day. When electricity and magnetism were first discovered, there were numerous attempts to apply them directly to explanations of animal bodies, but many of these early models were discarded later. In the twentieth century, animal processes are often conceptualized as computer entities, such as nervous system as a computational network. Likewise, animal brains are observed for the purpose of building the artificial intelligence. Mathematical models in biology developed from simple measurements of weight, length, and proportion to those incorporating calculus, differential equations, statistics, computational science, and other areas of modern mathematics.

Animal Motility, Field Perception, and Gradients

Animals can move under their own power. Animals movement in response to external stimuli or gradients of stimuli is called taxis. In calculus, the gradient is a vector field; its vectors point in the direction of the greatest rate of increase in a variable and have the magnitude equal to that rate. Depending on the nature of the variable in the gradient, animals or animal cells can exhibit different types of taxis, such as thermotaxis along temperature gradients or phototaxis along light gradients. Mathematical models of taxis are based on calculus, differential equations, and statistics.

Chemotaxis is the movement along the gradient in a chemical substance. Animal cells may have multiple chemical receptors around their boundaries, allowing the cell to determine the direction of chemical gradient vectors. Animal cells can move toward chemoattractors, such as immune cells arriving where they need to be, or away from chemorepellents. The development of animal embryos involves the movement of cells and is regulated by gradients in signal chemicals. Sperm movement occurs because of chemotaxis and thermotaxis.

Magnetoperception (the ability to detect magnetic fields) is observed in migrating birds, sharks, rays, honeybees, and other animals. It is an important factor in regulating animal movement and navigation—for example, during bird migrations. Experiments and applications in magnetoperceptions usually involve attaching magnetic substances to animals and observing effects. For example, cows and deer grazing under power lines orient themselves differently. The mechanisms of magnetoperception continue to be actively investigated.

Animal Locomotion

The way animals move, in addition to being a matter of biological interest, is a source of engineering ideas. Until the twentieth century, the main source of data on animal movement was observation and, sometimes, experiments with animals or their body parts. Photography and videography added details to the observation. Animals may be equipped with miniature devices that track their positions in space, as well as the electric activity within muscles, the contraction of muscles, or the forces exerted by muscles. These devices allow the development of detailed models of animal bodies during movement.

Every type of locomotion has been modeled in physics, with a variety of relevant equations. There are three major types of terrestrial locomotion (movement on solid surfaces): legged movement, slithering, and rolling. Legged animals may have from two to 750 legs, with the geometry of leg and joint position defining posture, and the pattern and pace of leg use defining gaits. Snakes move by undulating in several patterns, such as sidewinding, or by lifting parts of their belly slightly off the ground, moving them forward relative to their ribs, and then pulling the body to them (rectilinear motion). These movements on land are described by kinematic equations, in water by hydrodynamic equations. Rolling animals, such as pangolins, can briefly achieve great speed, usually by forming a wheel or a ball out of their whole body and using gravity to escape predators.

Swimming is accomplished by body movement propulsion in fish, jet propulsion in mollusks, undulation in several types of animals, and limb movement in some birds and mammals. Jet propulsion requires relatively high energy but can provide animals with an occasional burst of speed. Models of swimming include such measures as buoyancy and are modeled with fluid dynamics and mechanics.

Gliding, soaring, and flying are energy-efficient ways of locomotion, and attract much interest in biomechanics and aerodynamics. Scientists study concepts like lift and drag as well as ratios of wing measurements such as loading (weight to area). Animals use different types of motion through the air, which are defined by a combination of timing and geometry. For example, falling with increased drag forces that prolongs the fall can be either parachuting (when the angle to earth is more than 45 degrees) or gliding (when the angle is less than 45 degrees). Gliding animals such as fish and squirrels have aerodynamic adaptations including streamlining. The variable glide ratio is the ratio between the horizontal and the vertical speed components (lift to drag). A flying squirrel has a glide ratio of about two, and a human in a glider windsuit modeled after gliding animals has a glide ratio of about two and a half. Soaring birds glide during parts of their flight.

The properties of winged flight in birds and bats depend on proportions of the animal's body. Wingspan is the distance between wingtips, and the mean wing chord is the average of the distances between the front and the back edge of the wing, found using calculus.

Aspect ratio of a wing is the ratio of wingspan to mean chord. Fast birds such as falcons have pointy short wings with high aspect ratio (narrow wings). Long wings with high aspect ratios such as the wings of albatrosses, on the other hand, can produce slow soaring and gliding flight. Wide, rounded wings with medium aspect ratios can be used for a variety of flight types, for example, in storks or sparrows.

Biophysicists first attempted to explain insect flight using bird flight mechanics. They found that the resulting forces were several times less than what would be needed to lift and to propel an insect. Current theories of insect flight are still controversial. The theories use computational differential equations to model effects such as vortexes created in front of wings. When wings flap with high enough frequency, such a vortex can provide significant additional suction force.

Relatively rare types of animal locomotion depend on surface tension and capillary forces for walking on the water surface, or moving faster over released liquid (Marangoni effect). These forces are studied in fluid dynamics and thermodynamics.

Researchers debate why the wheel, which provides several mechanical advantages in terrestrial locomotion, has never evolved in any animal. The relevant mathematical model is a graph measuring fitness of organisms to the environment, called fitness landscape. Fitness peaks are stable states, with genetic modifications meaning worse fitness. While wheel locomotion may be a fitness peak, it is surrounded by fitness valleys too deep to be crossed by evolutionary means.

Migrating Animals

Many animals migrate—periodically travelling among habitats—sometimes over long distances. Models of migration take into account the time of each leg of the journey as well as the full period of migration. These times can be synchronized with seasonal milestones, developmental stages in the life of each animal, and other natural events. Because migrations can take place across international boundaries, they can help promote international efforts in research and conservation. The Convention on the Conservation of Migratory Species of Wild Animals, for example, covers several endangered species of birds and fish, as well as migratory bats and turtles.

About a fifth of all bird species in the world migrate. Typically, birds migrate closer to the equator in winters, and farther from the equator in summers. Mathematical models of bird migration include the overall patterns for particular populations such as migration corridors as well as random events such as irruptions (large numbers of birds) migrating farther after population explosions. Biophysics involved in bird migration includes theories of energy efficiency, and various mechanical effects, such as wear on feathers that necessitates periodic molting synchronized with the migration period.

During migration, birds navigate by using the landscape clues they learn while young, orienting by the sun, or using magnetoperception. In some bird species, navigating is mostly a learned behavior; in others, it is mostly coded genetically. Sometimes the coding goes wrong, reversing the migration direction 180 degrees, thus causing birds to reverse-migrate in the opposite direction from the majority of their flock. Bird species that learn their migration routes from their elders, such as cranes, can be taught to use safer routes by following light aircraft of animal preservation specialists.

Shorter migration routes also exist. For example, many fish species rise to the water surface to feed at night—a type of diel vertical migration. Many fish species high in the food chain migrate to follow their prey, with varying times and lengths of migration journeys.

Because many insects are relatively short-lived, their migrations may involve multiple generations

Snakes move with slithering terrestrial locomotion by undulating in one of several patterns. (Photos.com)

The long wings of albatrosses produce slow soaring and gliding flight. (Photos.com)

The Physics of Winged Flight

The physics of winged flight in birds, bats, and extinct dinosaurs focuses on the balance of four forces: lift, drag, thrust, and gravity. Air moving around wings produces lift because of speed and pressure differences in the airflow on either side of the wing—a complex process still being studied in aerodynamics and modeled with systems of differential equations.

Drag comes from air resistance to the flying body, and by air turbulence created by wing movements. Newton's second and third laws of physics explain thrust (the force created when a wing flaps). The vector of thrust points in the direction opposite to where the wing is moving. In other words, the flapping wing propels the animal forward by by pushing against the air. The force of gravity is proportional to the mass of the animal.

being born along the route. In these cases, none of the individual insects travels the full migration route. Some migrating insects, such as locusts, swarm for the purpose of migration. A swarm can be modeled using a system of differential equations where pairs of individuals move closer if they are too far, move away if they are too close, and orient themselves toward the same direction. However, studies of insects, including locusts, show complex mechanisms that include chemoregulation, physiological change in response to overcrowding (measured in contacts per unit of time), emission, and responsiveness to sounds and other variables involved in swarming.

Herds of animals, schools of fish, and flocks of birds can be modeled as groups of particles, with interactions among individuals determined by differential equations with some fixed and some random parameters to account for individual behavior variations. Such mathematical models (called "interacting particle models") can describe flock behavior or predict school migration routes. To observe animal migration, researchers use tracking devices, satellite observation, and echolocation for marine species.

Food Webs

Food webs and food chains map food relationships in ecosystems. The key measurement of the position within the food web is called "trophic level." Autotrophs (producers) are at trophism level one. Autotrophs are organisms that do not consume other organisms or carbon produced by them, and therefore are not animals. Two mechanisms of autotrophism are photosynthesis in plants, and chemosynthesis in archae and bacteria. The first organisms to evolve on Earth used chemosynthesis. A third mechanism, radiotrophy, is being researched in fungi in high-radiation areas. All food chains within all food webs on Earth start with level one autotrophs. Predator species that no other species predate upon are called apex predators.

More specifically, classes of organisms are named according to the flow chart with three branchings. The first branching determines the source of energy, either light (photo-) or chemical (chemo-). The second branching determines the source of extra electrons in reduction-oxidation reactions, either organic (-organo-) or inorganic (-litho-). The third branching defines the source of carbon, either organic (-heterotroph) or carbon dioxide (-autotrophs). For example, fungi are chemoorganotrophic. All eight combinations resulting from these three branchings exist in nature. Heterotrophic organisms that break down other dead organisms into simpler organic or inorganic compounds are called decomposers. Consumer organisms use other living organisms as their source of energy. Simplistically, the second trophic level comprises primary consumers that eat plants (herbivores) or chemosynthesizing creatures. The third trophic level, secondary consumers or predators, consists of animals that eat primary consumers. Animals that eat those at the third trophic level are said to have the fourth trophic level, and so on. However, most existing animal species obtain energy from several sources. For example, foxes eat rabbits and berries; chickens eat grains and insects.

To address the complexity of food chains, the trophic level of an animal is determined by the formula of adding all products of levels of its food by the fraction of that food in the animal's diet, and adding 1. For example, if a chicken's diet consists of 30% worms (level 2) and 70% grain (level 1), its trophic level is equal to

$$0.3(2) + 0.7(1) + 1 = 2.3.$$

Statistical analysis is used to determine the mean trophic level of a species in a particular ecosystem.

Changes in any part of the food web affect all other parts. For example, the effect of introducing predators that reduce the numbers of the prey and cause abundance in the next trophic level down is an example of a "trophic cascade event." The ability of an ecosystem to withstand disturbances is measured by an index called ascendency, and is derived by formulas from the information theory field of mathematics. Variables in ascendency formulas include both the amounts of energy and matter circulated within an ecosystem and the information shared among members of the system. Low ascendency values make ecosystems internally unstable; high ascendency values make ecosystems oversensitive to external disturbances. Ascendency values corresponding to stable systems are called "the window of vitality."

Ascendency is an example of using multiple indices and metrics to model, evaluate, and predict changes in food webs. For example, consider energy or biomass transfer from one feeding level to the next feeding level. The efficiency of this transfer is a measure of an ecosystem called ecological efficiency. For example, in a food chain that consists of four levels, with mean ecological efficiency of 1/10, the apex predator has the ecological efficiency of converting sunlight into its biomass of

$$\left(\frac{1}{10}\right)^4 = 0.0001.$$

Ecological efficiency restricts the number of possible trophic levels.

Fantastic Animals, Hybrids, and Genetic Chimeras

A variety of cultures describe fantastic animals or humanoids with animal traits. These animals—especially those invented before the nineteenth century—are used in mathematics education to help students understand concepts related to combinatorics because they are made by combining parts of existing animals. For example, ancient Greeks invented a chimera that had the body of a lion, the heads of a goat and a lion, and a snake for the tail. In genetics, chimeras are animals that have genetic material from more than one zygote—from four or more parents. Chimeras of different animals of the same species happen naturally when several eggs in one female are fertilized by sperm from different males and then fused. They may also happen artificially, in which case different animals species can be used. For example, a goat-sheep chimera called "geep" was first produced in the 1970s.

Hybrid animals are different from chimeras in that they have two parents, but the parents are of different species. Hybridization has been recognized and used for millennia. For example, humans have produced large populations of mules since ancient times. The mathematics of hybrids involves tracking the amount of genetic material from each species through generations, and calculating the probabilities of achieving particular traits in offspring. For example, a single-cross hybrid has 50% genetic material from either line of parents. Crossing such hybrids with the line of one of the parents (called backcrossing) produces hybrids with roughly 75% genetic material from that parent's species—averaged across a species, as individuals will have either pure or half-and-half genetic material.

Symmetry and Fractals

Most animal bodies exhibit either rotational (radial) or reflection symmetry. Animals with bilateral reflection symmetry (having a plane separating bodies into roughly reflected halves) form the taxon Bilateria. Observation of symmetry is a major tool of evolutionary theory. For example, it is hypothesized that all Bilateria animals evolved from a common ancestor species, Urbilaterian, that lived around six hundred million years ago. This makes Bilateria a clade (a group of animals that come from a common ancestor). Bilaterians have the front end with the mouth and the back end with the anus, defined by the plane of symmetry.

Rotationally symmetric animals such as sea anemones and sea stars usually have the mouth on the axis of the symmetry. When animals have a certain number of body regions positioned around the axis symmetrically, they are called by the number of regions. For example, five-armed stars exhibit pentamerism, and many coral polyps exhibit hexamerism, or six-part rotational symmetry.

Combinations of reflections, rotations, and translations can produce repeated geometric patterns called tessellations or wallpaper groups in plane and crystallographic groups in space. There are 17 types of wallpaper groups and 230 types of crystallographic groups described by the area of mathematics called group theory. Wallpaper and crystallographic groups can be found in colonies of animals such as corals or in arrangements of animal body parts such as fish scales.

Fractals are shapes that can be split into parts that are copies of the whole. Fractals frequently occur in the living nature. For example, feathers are fractal-like structures of the tree type, with three or four levels. Nervous systems and lungs of mammals are also tree-type fractals. Beyond the literal meaning as a geometric shape, the idea of a fractal as a self-repeating structure is applied to many areas related to animals to describe patterns within systems behavior, evolution, migration, and development.

Further Reading

Adam, John. *A Mathematical Nature Walk*. Princeton, NJ: Princeton University Press, 2009.

Ahlborn, Boye. *Zoological Physics: Quantitative Models of Body Design, Actions, and Physical Limitations of Animals*. New York: Springer, 2004.

Ball, Philip. *The Self-Made Tapestry: Pattern Formation in Nature*. New York: Oxford University Press, 2001.

Cheng, Ken, and Nora Newcombe. "Geometry, Features, and Orientation in Vertebrate Animals: A Pictorial Review." Animal Spatial Cognition: Comparative, Neural & Computational Approaches, 2006. http://www.pigeon.psy.tufts.edu/asc/Cheng/Default.htm.

Devlin, Keith. *The Math Instinct: Why You're a Mathematical Genius (Along With Lobsters, Birds, Cats, and Dogs)*. New York: Basic Books, 2005.

Murray, James. "How the Leopard Gets Its Spots." *Scientific American* 258, no. 3 (1988).

Noonan, Diana. *Animal Investigations: Collecting Data*. Huntington Beach, CA: Teacher Created Materials, 2008.

Maria Droujkova

Astronomy

Category: Space, Time, and Distance.
Fields of Study: Geometry; Measurement; Number and Operations; Representations.
Summary: Mathematics is used in astronomy to measure and model celestial bodies.

Astronomy is the science that deals with celestial objects. It is divided into two disciplines: positional astronomy (or "astrometry"), which deals with the positions and movements of celestial objects; and astrophysics, which deals with their chemical and physical properties.

Positional astronomy began as a practical science. The first astronomers, before the invention of writing, dealt with such questions as the proper time of the year to plant crops and the proper dates for religious festivals. As their understanding improved over the ages, astronomers tackled other practical problems such as how to predict eclipses, how to tell time within a day, and how to navigate at sea.

Ancient people could take simple observations of the sun and moon and observe the patterns they made. From there it was a short leap to predicting future patterns. They would first record (or, before writing, memorize) the observations, and then perform a mathematical analysis—even if the analysis were nothing more than counting (for example, discovering there were about 365 days between winter solstices).

A much more sophisticated accomplishment was working out the complicated cycles on which lunar and solar eclipses occurred. An eclipse can be terrifying for a people who are not expecting it. If astronomers (many of whom doubled as priests) could predict eclipses, they could warn people in advance and reduce the collective fear.

A number of ancient peoples, including Mayans, Chinese, and Babylonians, developed elaborate calendar systems and tracked the movements of the planets. The Chinese constructed star charts, kept records starting possibly as early as 4000 B.C.E., and developed astronomical instruments. The Babylonians mapped constellations and introduced 60-minute hours and 60-second minutes. Both the Chinese and the Babylonians were able to predict eclipses. By 2500 b.c.e, Egyptians had measured star positions well enough to orient the pyramids to face celestial north. Polynesians traveled throughout the Pacific Ocean using stars as navigational aids.

The Greeks

The ancient Greeks effectively applied mathematics to astronomy. Eratosthenes (c. 200 B.C.E.) used geometry to calculate the size of Earth. Hipparchus (c. 161–126 B.C.E.) discovered the precession of the equinoxes and created the most accurate Greek tables of lunar motion. Like some other Greek astronomers, he held that Earth revolved around the sun. Claudius Ptolemy (c. 120–150 C.E.) combined observations from Hipparchus and others with his own observations to propose a model of how the solar system worked—assuming Earth was at the center. By using epicycles (circles revolving on circles), he produced what was by far the best model of the heavens until Nicolaus Copernicus.

The Greeks did not only conduct astronomical calculations by hand but used a computer as well. Though not much is currently known about it, a mechanical analog computer was built somewhere in the Greek world about 100 B.C.E., called the "Antikythera mechanism" after the place it was found. This remarkably sophisticated computer was able to show both solar and lunar calendars, track the complicated path of the moon using Hipparchus's results, and predict eclipses for years into the future.

The Renaissance

During the Middle Ages, Arabs, Persians, and Jews, as well as European Christians (after c. 1000 C.E.), continued the work of the Greeks, including making new tables of planetary positions to update Ptolemy's, and keeping track of the precession of the equinoxes. In 1543, Copernicus's book on the solar system was published. Through a mathematical analysis of Ptolemy's work and later observations, Copernicus showed that a system in which the sun was the center of the solar system led to simpler and more accurate analysis than Ptolemy's.

(Photos.com)

Lunar Calendars

Some cultures found it easier to use the moon instead of the sun to tell time. A quick glance shows the phase of the moon, while observing the sun takes careful measurement. There is not an even number of moon cycles in a solar year—making the exclusive use of moon calendars difficult—yet the Babylonians (and others) discovered that there was a 19-year cycle.

In 19 solar years (assuming 365.25 days per year), there are 6939.75 days. The moon takes 29.5306 days to cycle from one new moon to the next. In 235 lunar months, there are 6939.69 days. A lunar calendar based on this cycle has 12 months (six with 29 days and six with 30 days—figuring a lunar month of 29.5 days), adding up to 354 days. Seven times in the 19-year cycle, there is a leap year, with an extra month of 30 days added in—making the leap year 384 days.

$$(12 \times 354) + (7 \times 384) = 6936 \text{ days.}$$

Since the lunar month is slightly more than 29.5 days, a total of four days have to be added during the 19-year cycle, giving 6940 days. This lunar calendar, originated by the Babylonians and refined over the centuries, is still in use today—it sets the dates of Easter and of all Jewish holidays.

Johannes Kepler used Tycho Brahe's careful naked eye observations of the planets to show that Mars went around the sun in an ellipse, not a circle as the Greeks had assumed. Kepler stated his three laws, which relate the speed of a planet to the shape of its orbit, but he could not explain why these laws worked. Isaac Newton was the first to explain Kepler's laws. He was able to show that any object affected by gravity would move in one of the conic sections: Kepler's ellipse, a line, a circle, a parabola, or a hyperbola. The one exception was the planet Uranus, which did not follow its Newtonian orbit.

It was not until the 1800s that Urbain Leverrier, in France, and John Couch Adams, in England, (unknown to each other) made the assumption that the discrepancies were because of the gravitational pull of an unknown planet. The planet Neptune was discovered in 1846 using Leverrier's prediction. Neptune was found by the consideration of the three components, P_x, P_y, and P_z, of Neptune's position and the three components, V_x, V_y, and V_z, of Neptune's velocity.

Until 1821, Uranus was moving faster in its orbit than expected—more than 4 planetary diameters ahead of its predicted position. After 1821, Uranus moved slower than expected. Obviously, Uranus moved past Neptune around 1821. If one adjusted the coordinate system so that $P_x = 0$ was Uranus's position in its orbit in 1821 and examined how far Uranus was pulled above or below its expected orbit, then one can tell whether Neptune was above or below Uranus in 1821, which gives P_y, and also whether Neptune was moving up or down, which gives V_y. If we have P_z, which represents Neptune's distance from the sun in 1821, then Kepler's laws can be used to find the two remaining parameters: V_x and V_z. Leverrier and Adams used a shortcut to find P_z. Both used Titius-Bode's law, an empirical formula, to predict the next planet beyond Uranus to be 38.8 times Earth's distance from the sun. These predictions were good enough to find Neptune, although Neptune is only 30.1 times Earth's distance from the sun.

Leverrier later examined the orbit of Mercury and found a discrepancy of 43 seconds of arc (which sounds small but is twice the discrepancies used to find Neptune). He computed the orbit of a hypothetical planet, called "Vulcan," which would explain this 43-second variation. Vulcan has never been found, and Einstein's general theory of relativity also explains this discrepancy.

Parallax

The ancient Greeks made attempts using parallax (the difference in the angle to a distant body measured from two different locations, also called triangulation) to find the size of the solar system. Being restricted to naked-eye observations, their results were inaccurate. Using telescopes, a much more accurate measurement was made in 1761 in which observers scattered across Earth found the parallax of Venus when it passed in front of the sun. The observations gave a value of 95.25 million miles from Earth to the sun (the modern estimate is just under 93 million miles). A much more difficult problem was to find the distances of stars by their parallax when viewed from opposite sides of Earth's orbit, first accomplished by Friedrich Bessel in the 1830s. Is space Euclidean or non-Euclidean? If measurably non-Euclidean, this would show up in stellar parallax measurements. No such effect has yet been observed, so one can say—except for relativistic considerations—that space is Euclidean for hundreds of light-years from Earth.

Astrophysics

Astrophysical questions date to the ancient Chinese, who discovered sunspots, and Hipparchus (c. 190–120 b.c.e), who worked on the magnitude (or brightness) of stars. His magnitudes, much refined, are still in use today. However, astrophysics as a discipline can be said to have started with Joseph von Fraunhofer, who in 1815 devised a spectroscope and catalogued the various lines (known as the Fraunhofer lines) that can be seen in the solar spectrum. In the 1850s, Gustav Kirchhoff and Robert Bunsen determined that these lines belonged to different chemical elements. Thus, by examining the spectrum of a star, its chemical composition can be determined. In addition, it was discovered that magnetic fields caused broadening and splitting of Fraunhofer lines, allowing the magnetic fields of stars to also be investigated.

Over the course of the twentieth century, astrophysicists went from studying the spectrum of visible light to studying every frequency of electromagnetic waves—from gamma rays to radio waves. There is now no known radiation from a star that is not being used to help find answers to the questions of what stars are, and how they operate.

Further Reading

Freeth, Tony. "Decoding an Ancient Computer." *Scientific American* 301, no. 6 (December 2009).

Gould Jr., Benjamin Althrop. *Report to the Smithsonian Institution on the History of the Discovery of Neptune*. Washington DC: Smithsonian, 1850.

Hester, Jeff, Bradford Smith, George Blumenthal, and Laura Kay. *21st Century Astronomy*. New York: W. W. Norton, 2010.

James Landau

Bees

Category: Weather, Nature, and Environment.
Fields of Study: Algebra, Geometry; Representations.
Summary: Geometry explains why honeycombs are made of hexagonal cells, while bee movement patterns communicate information visually.

Honeycombs are remarkable for their beauty, precision, and symmetry. The honeycomb corresponds to a mathematical concept known as a "tiling of the plane." That bees use regular hexagons for this tiling (built to a remarkable level of precision) has fascinated human beings throughout history. At the end of the twentieth century, mathematician Thomas Hales rigorously proved a long-standing conjecture that fully justifies to humans what the bees have apparently known all along: the most efficient way to repeatedly enclose a fixed amount of storage space is to use regular hexagons to form the boundaries.

Honeycomb: How to Choose a Cell

Bees use honeycomb cells for storage. It takes work and material (wax) to create the boundary of each cell, so the bees want cells with as little boundary (perimeter) as possible, given that each cell should enclose a certain amount of storage (area). If a bee only needed to make one cell to store honey, it would likely use a shape other than a regular hexagon. For instance, a regular octagon holding the same area has less perimeter; a regular decagon will have less perimeter still. The more sides a polygon has, the smaller the perimeter will be, with the circle having the smallest perimeter-to-area ratio. That

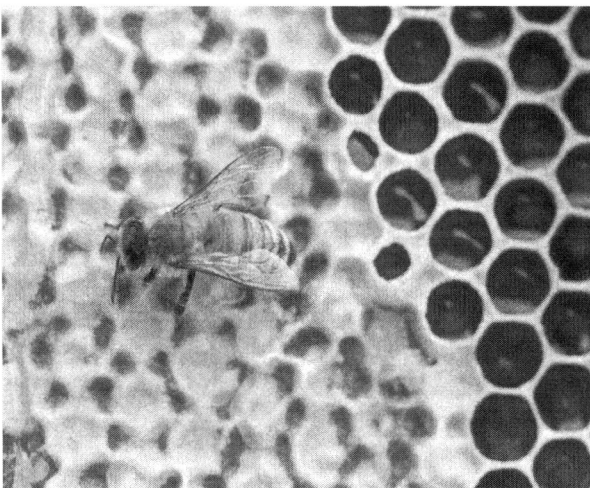

The most efficient way to repeatedly enclose a fixed amount of storage space is to use regular hexagons. (Photos.com)

a circle is the least-perimeter shape to enclose a given area is a famous problem that goes back to the wonderful tale of Queen Dido of Tyre.

For example, suppose a bee wanted to enclose one square unit of area. The square that accomplishes this has a perimeter of 4. If the bee used an equilateral triangle instead, the necessary perimeter is larger, about 4.56. But the regular hexagon's perimeter is smaller, at just over 3.72. The pattern of increasing the number of sides leading to a lower perimeter holds for all whole numbers $n > 2$, and every such regular polygon enclosing one unit of area has greater perimeter than a circle holding the same area. The circle that encloses one unit of area has a perimeter of approximately 3.54.

Tilings: Fitting the Cells Together

Bees do not need just one cell; they need many consecutive cells in which to place their honey, and therefore essentially have to create a "tiling" (a pattern involving polygons that will completely cover their work space without overlapping while leaving no space unused). Circular cells simply don't fit together as well because there are gaps between consecutive circles.

Many different kinds of floors and ceilings are tiled—usually with congruent squares or rectangles. Why don't bees use square cells in their honeycomb, rather than hexagons? Or equilateral triangles? It turns out that equilateral triangles, squares, and regular hexagons can all be used to tile the plane, as shown in the figures below. Bees choose hexagons from among these three options since a regular hexagon of unit area uses less perimeter (wax) than does a square or equilateral triangle; the hexagon is a more efficient choice (see Figure 2).

Figure 2.

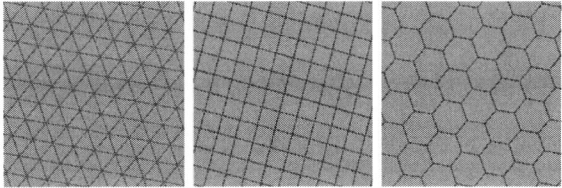

So why not use regular octagons? Here it is not the efficiency of the individual cell that governs the choice but rather the overall packing of them: regular octagons cannot be used to tile the plane.

To understand why triangles, squares, and hexagons tile the plane, but octagons do not, observe that in a regular polygon with n sides, the sum of its interior angles is $180(n-2)$ degrees, and each of its n individual interior angles has the measure

$$\frac{180(n-2)}{n}.$$

For instance, with the square, each interior angle has the measure

$$\frac{180(4-2)}{4} = 90 \text{ degrees.}$$

Four squares arranged at a single vertex fit together perfectly, creating a full 360 degrees around the shared corner. Likewise, six equilateral triangles (each having 60-degree angles) can fit together perfectly for a full 360 degrees, as can three regular hexagons with their 120-degree interior angles.

But for the octagon with $n = 8$, each interior angle has the measure of 135 degrees. Three octagons put together at a shared vertex would have $135 \times 3 = 405$ degrees, which is simply impossible—as would be attempting to only have two octagons meet at a single vertex. Regardless of the number of sides of the regular polygon, the measure of the polygon's interior angle will need to divide evenly into 360 degrees. This forces

$$\frac{2n}{n-2}$$

to be an integer, and the only values of n for which that is true are $n = 3, 4,$ and 6: triangle, square, and hexagon! That the only ways to tile a flat surface using congruent regular polygons are with triangles, squares, or hexagons is a result often taught in high school geometry courses.

Irregular and Non-Polygonal Tilings

Since the time of the ancient Greeks, mathematicians conjectured that among all the ways to tile the plane so that each tile encloses just one unit of area, the way that uses the least perimeter is the tiling that uses all regular hexagons. This conjecture is much harder than it sounds to prove: one must consider irregular polygons (with sides of different lengths), as well as the possibility that the sides of some tiles might be curved. The first possibility is not too difficult to eliminate. For instance, it is straightforward to show that a regular hexagon with all sides of equal length will use less perimeter than any other hexagon to enclose the same area.

But the second possibility—using non-polygonal shapes—proved to be much, much more challenging. In this situation, one must consider the possibility of a shape that bows out on one side and, to fit into a tiling, bows in on another. Obviously, the part that bows out picks up area, while the part that bows in loses area. In 1999, mathematician Thomas Hales proved that any advantage that comes from a side of the tile bowing out is more than cancelled out by the disadvantage that follows from another side having to bow in. Thus, the ideal tile is one that has no bulges: a polygon!

What Professor Hales proved is essentially what the bees knew all along: of all possible tilings, the one using regular hexagons is the most efficient way to enclose cells of the same area.

Other Mathematical Aspects of Bees

Another way that mathematics relates to bees is when mathematicians work with bee researchers to solve problems such as those related to viral disease infection and pollination. Mathematics is also used to model the ways in which bees communicate locations. When a bee finds a source of food, it returns to the hive and performs an elaborate dance that conveys the direction and distance from the hive. Ethologist Karl von Frisch was one of the first to explore the meaning of the honeybee dance, and he won a Nobel Prize for his work. The angle that the bee dances expresses the direction. For example, if a bee dances in a straight line toward the upper part of the hive, then the flowers are located in the direction of the sun. The bee also takes into account the fact that the sun moves; the angle it describes inside the hive changes as the sun does. The duration of the dance and the number of vibrations give the exact distance. Other features of the dance remained unexplained until Barbara Shipman theorized that the honeybee's complex choreography is a projection of a six-dimensional space, and she was able to use this representation to reproduce the entire bee dance in all its parts and variations. To her, this implies that bees can sense the quantum world, although some researchers dispute her conclusions.

Further Reading
Austin, David. "Penrose Tiles Talk Across Miles." http://www.ams.org/featurecolumn/archive/penrose.html.
Frank, Adam. "Quantum Honeybees." *Discover* 18 (1997).
Hirsch, Christian R., et al. *Core-Plus Mathematics: Contemporary Mathematics in Context, Course 1.* 2nd Ed. Columbus, OH: Glencoe/McGraw-Hill, 2008.
Morgan, Frank. "Hales Proves Honeycomb Conjecture." http://www.maa.org/features/mathchat/mathchat_6_17_99.html.
Peterson, Ivars. "The Honeycomb Conjecture: Proving Mathematically That Honeybee Constructors Are on the Right Track." *Science News* 156, no. 4 (July 24, 1999). http://www.sciencenews.org/sn_arc99/7_24_99/bob2.htm.
Serra, Michael. *Discovering Geometry: An Inductive Approach.* 4th ed. Emeryville, CA: Key Curriculum Press, 2008.
University of Guelph. "U of G Using Math to Study Bees." http://www.uoguelph.ca/news/2009/12/u_of_g_using_ma.html.

Matt Boelkins

Black Holes

Category: Space, Time, and Distance.
Fields of Study: Geometry; Measurement; Number and Operations; Representations.
Summary: Black holes were implied by Einstein's general relativity and have challenged physicists' theories since.

A black hole is a finite region of space during a period of time (called space-time) subject to a singularity caused by a large concentration of mass in its interior. This massive object generates a gravitational field so powerful that atoms are compacted in super-high densities, which in turn increases the gravitational pull. A singularity is created in space because no particle of matter, not even light photons, can escape from that region. Hence the name: a black hole is an invisible region because it does not reflect any light (all light is absorbed). Many aspects of black holes can be described and studied using algebraic and geometric concepts, but the existence of black holes is still under debate. For example, Australian mathematician Stephen Crothers argues that black holes are inconsistent with general relativity and critiques the mathematics used by others to demonstrate their existence. It is believed that black holes originate when stars runs out of gas needed to maintain their temperature, causing a decrease in volume. As volume decreases, the proximity of particles increases the gravitational pull in a positive feedback loop; as particles get closer, the gravitational force keeps increasing. This compaction process continues until a singularity, called the "event horizon," is created. The event horizon is defined as a boundary in space and time beyond which events cannot affect an outside observer. The event horizon separates the black hole region from the rest of the universe and is the boundary of space from which no particle can leave, including light.

The singularity caused by a black hole is considered as a curvature in space-time. This curvature is explored by Albert Einstein's general relativity theory, which predicted the existence of black holes—though Einstein himself did not believe in them. In the 1970s, Stephen Hawking, George Ellis, and Roger Penrose proved several important theorems on the occurrence and geometry of black holes. Previously, in 1963, Roy Kerr had shown that black holes in a space-time have an almost-spherical geometry determined by three parameters: their mass, their total electric charge, and angular momentum.

It is believed that at the center of most galaxies, including the Milky Way, there are supermassive black holes. The existence of black holes is supported by astronomical observations, in particular through the emission of X-rays. Some black hole candidates have been identified experimentally using observations and data. There are different types of black holes, such as rotating black holes and stationary black holes, and these are described by using various metrics in physics and differential geometry.

Origins of Human Awareness of Black Holes

The concept of a body so dense that even light could not escape was described in a paper submitted in 1783 to the Royal Society by an English geologist named John Michell. By then, Isaac Newton's theory of gravitation and the concept of escape velocity were well known. Michell computed that a body with a radius 500 times that of the sun and the same density, would, on its surface, have an escape velocity equal to that of light and would therefore be invisible.

In 1796, the French mathematician Pierre-Simon Laplace explained in the first two editions of his book *Exposition du Système du Monde* the same idea; however, the concept that light was a wave without mass and therefore unaffected by gravitation was prevalent in the nineteenth century, and Laplace discarded the idea in later editions.

In 1915, Einstein developed his general relativity theory, and showed that light was influenced by the gravitational interaction. A few months later, Karl Schwarzschild found a solution to Einstein's equations, where a heavy body would absorb the light. We now know that the Schwarzschild radius is the radius of the event horizon of a black hole that will not turn, but this was not well understood at the time. Schwarzschild himself thought it was just a mathematical solution, not physical. In 1930, Subrahmanyan Chandrasekhar showed that any star with a critical mass (now known as the Chandrasekhar limit) and that does not emit radiation would collapse under its own gravity. However, Arthur Eddington opposed the idea that the star would reach a size zero, implying a naked singularity of matter; instead, the black hole should have some-

An artist's concept chronicles a time lapse from left to right of an intact sun-like star (left) coming too close to a black hole (right) and its self-gravity becoming overwhelmed by a black hole's gravity. (National Aeronautics and Space Administration)

thing that will inevitably put a stop to collapse, an idea adopted by most scientists.

In 1939, Robert Oppenheimer predicted that a massive star could suffer a gravitational collapse and therefore black holes might be formed in nature. This theory did not receive much attention until the 1960s because after World War II he was more interested in what was happening at the atomic scale.

In 1967, Stephen Hawking and Roger Penrose proved that black holes are solutions to Einstein's equations and that in certain cases the creation of a black hole is the inevitable consequence of a star aging. The black hole idea gained force with the scientific and experimental advances that led to the discovery of pulsars. Soon after, in 1969, John Wheeler coined the term "black hole" during a meeting of cosmologists in New York, to designate what was formerly called "star in gravitational collapse completely."

The Entropy of Black Holes

The mathematical tools used to model black holes use fundamental laws of physics, particularly relativity and thermodynamics. According to initial theories by Stephen Hawking, black holes violate the second law of thermodynamics (the entropy, or disorder, of isolated systems tend to increase over time), which led to speculations about travel in space-time wormholes (tunnels that would allow time travel or fast travel over very long distances). Hawking has recanted his original theory and has admitted that the entropy of the matter is kept inside a black hole. According to Hawking, despite the physical impossibility of escape from a black hole, it may end up evaporating by constant leakage of X-ray energy that escapes the event horizon, called Hawking radiation. According to this model, black holes have intrinsic gravitational entropy, which implies that gravity introduces an additional level of unpredictability over the quantum uncertainty. It appears, based on the current theoretical and experimental capacity, as if nature took decisions by chance or, more generally, far from precise laws.

The hypothesis that a black hole contains entropy and, furthermore, it is finite, required to be consistent with such holes emitting thermal radiation, at first seems contradictory. The explanation is that the radiation escapes the black hole in such a way that an external observer knows only the mass, angular momentum, and electric charge. This means that all combinations or configurations of radiation of particles having energy, angular momentum, and electric charge are equally likely. Physicists such as Jacob D. Bekenstein have been linked to black hole entropy and information theory.

Further Reading

Hawking, Stephen. *A Brief History of Time*. New York: Bantam Books, 1998.

———. *The Universe in a Nutshell*. New York: Bantam Books, 2001.

Poisson, Eric. *A Relativist's Toolkit: The Mathematics of Black-Hole Mechanics*. Cambridge, England: Cambridge University Press, 2007.

Sagan, Carl. *Cosmos*. New York: Random House, 2002.

Juan B. Gutierrez

Brain

Category: Medicine and Health.
Fields of study: All.
Summary: The brain is studied through models and through algorithm-dependent medical technology. The neurology of mathematical thought is a vibrant field.

The applications of mathematics to the study and understanding of the brain have been varied and widespread. They include models to predict the start of seizures using dynamical systems; maps of the brain using projective, hyperbolic, or other geometries, as well as graph theory; applications of morphometrics, which is the statistical study of shapes, to schizophrenic brains; and dynamic simulations and visualizations of electrochemical activity in neurons. Other models are used to study how electrical signals propagate along nerve cells and the way in which electrical discharges in nerve cells tend to synchronize and form waves. Medical technology used in brain treatment and studies uses mathematical algorithms; for example, to create and process computer-generated images of brain cells, as well as to measure functions like blood flow, glucose consumption, and electrical activity. Mathematics is also important in modern medical devices that involve nerve fibers within or leading to the brain, such as cochlear implants. How mathematical thought arises in the brain—from arithmetic to abstract thinking—is also of great interest. Mathematics and the Brain was the theme of Mathematics Awareness Month in 2007.

Brain Composition and Structure

Before proceeding with some applications of mathematics in the study of the brain, it is important to have an idea of brain composition and structure. In humans, this complex organ consists of perhaps 100 billion nerve cells (or neurons), with roughly a total of 100 trillion connections between neurons (or synapses). Although some nerve cells do regenerate, and new connections between nerve cells are made, overall these numbers tend to decline after birth. Even with advances in computer processing and storage, the sheer number of neurons and connections hints at the enormous scope of the problem inherent in understanding the brain. By comparison, the nematode *Caenorhabditis elegans* has 959 cells in the entire organism, 302 being nerve cells, which result in over 5000 connections between neurons. Even for something of this vastly smaller scale, the nematode's neural connections were initially mapped after more than 10 years of effort by the mid-1980s, and earned a Nobel Prize for Sydney Brenner—who famously called *C. elegans* "nature's gift to science." Those results have since been updated.

A single neuron generally consists of: a main cell body (or soma); many filamentous dendrites, which are where signals from other neurons are usually received; and a single axon, which typically communicates to the dendrites of other neurons. The electrical voltage across the neuron's cell membrane varies as the concentrations of calcium, sodium, potassium, and chloride ions fluctuate, producing a fluctuating electrical signal. The electrical signal is transferred from one neuron's axon to another's dendrite across a gap known as a "synapse." Some synapses, known as "electrical synapses," involve a direct channel that connects the two cells' cytoplasm and allows for very fast electrical transmission. By contrast, chemical synapses involve molecules known as "neurotransmitters," which mediate signal transmission. The human brain utilizes more than 100 types of neurotransmitters. However, just two of these types arise at the vast majority of synapses; namely, glutamate and gamma aminobutyric acid. In addition to neurons, glial cells serve various support functions for neurons. One important function of special kinds of glial cells—namely, special kinds of oli-

godendrocytes—is the myelinization of axons. Myelin, a fatty substance, essentially electrically insulates neurons. Because they are pinkish white, and white when stored in formaldehyde, bundles of myelinized axons make up what is known as "white matter" in the brain. On the other hand, "grey matter," as seen on the surface of the cerebral cortex in a typical brain slice image, comprises of the soma, dendrites, and other kinds of glial cells, such as astrocytes. While *C. Elegans* has fewer than 60 glial cells, the human brain likely has at least as many glial cells as neurons, although the ratio varies widely in different brain regions.

Applications of Neural Networks

How neurons collectively convey information is also of much interest to researchers. Interestingly, attempts to model so-called "artificial neural networks" have led to highly useful algorithms used in many areas of mathematics, science, and engineering in their own right, having nothing to do with the study of the brain. Computers are often "trained" with data sets using such neural networks to help process data. Neural networks can be found in software used in fields as varied as financial analysis and fraud detection, robotics, handwriting analysis, and voice recognition. As another example, much mathematics is used in processing and analyzing the enormous amount of neuron image data, and neural network algorithms are now being used to help automate that processing to help computers track neural connectivity.

Brain Mapping and Study

Mathematics also has been used to help in producing accurate maps of the cortex of various parts of the brain. The extensive folding in the human brain in the cerebral cortex, which produces peaks or ridges (or *gyri*) and valleys or furrows (or *sulchi*) makes it difficult to compare two different brain surfaces. A calculus-based geometry is used to find effective maps. As another example of an application, mathematics is used extensively in devices such as cochlear implants, useful to deaf individuals who still have a functioning auditory nerve. In humans, as many as 30,000 individual nerve cells in the inner ear pass through the auditory nerve to the brain. Different sound frequencies innervate different nerve cells; roughly speaking, lower frequencies innervate nerve cells in the basilar membrane closer to the beginning of the cochlea, as opposed to higher frequencies innervating cells further along. But the precise mapping of which cells are affected by which frequencies follows a logarithmic mathematical pattern, as a function of distance in the cochlea. Using various radio signal technologies, external sounds are transmitted to a receiver in the inner ear, which connects to implanted electrodes for nerve innervation. There, mathematics is used in the computational processing to convert the received frequencies into the appropriate electrical innervations, so that only certain nerve cells are stimulated for certain frequencies.

Examples of mathematics applied to the study of the brain abound in the five-year, National Institutes of Health–funded Human Connectome Project. This project, somewhat analogous to the Human Genome Project, was funded in 2010 for approximately $40 million. Mapping all the connections between neurons in the human brain in a meaningful way is the goal of the Connectome project. One component involves constructing connection data from 1200 individuals, including numerous twins. Developing effective ways to collect the data set, as well as analyze the results, involves several areas of mathematics in crucial ways. First, instruments must be able to create high-resolution images of the brain tissue of living humans in a completely noninvasive way. Next, the enormous image data must yield to automated computer analysis that can determine the actual neural connections within the brain. Finally, the connection data set must be amenable to meaningful analysis by researchers

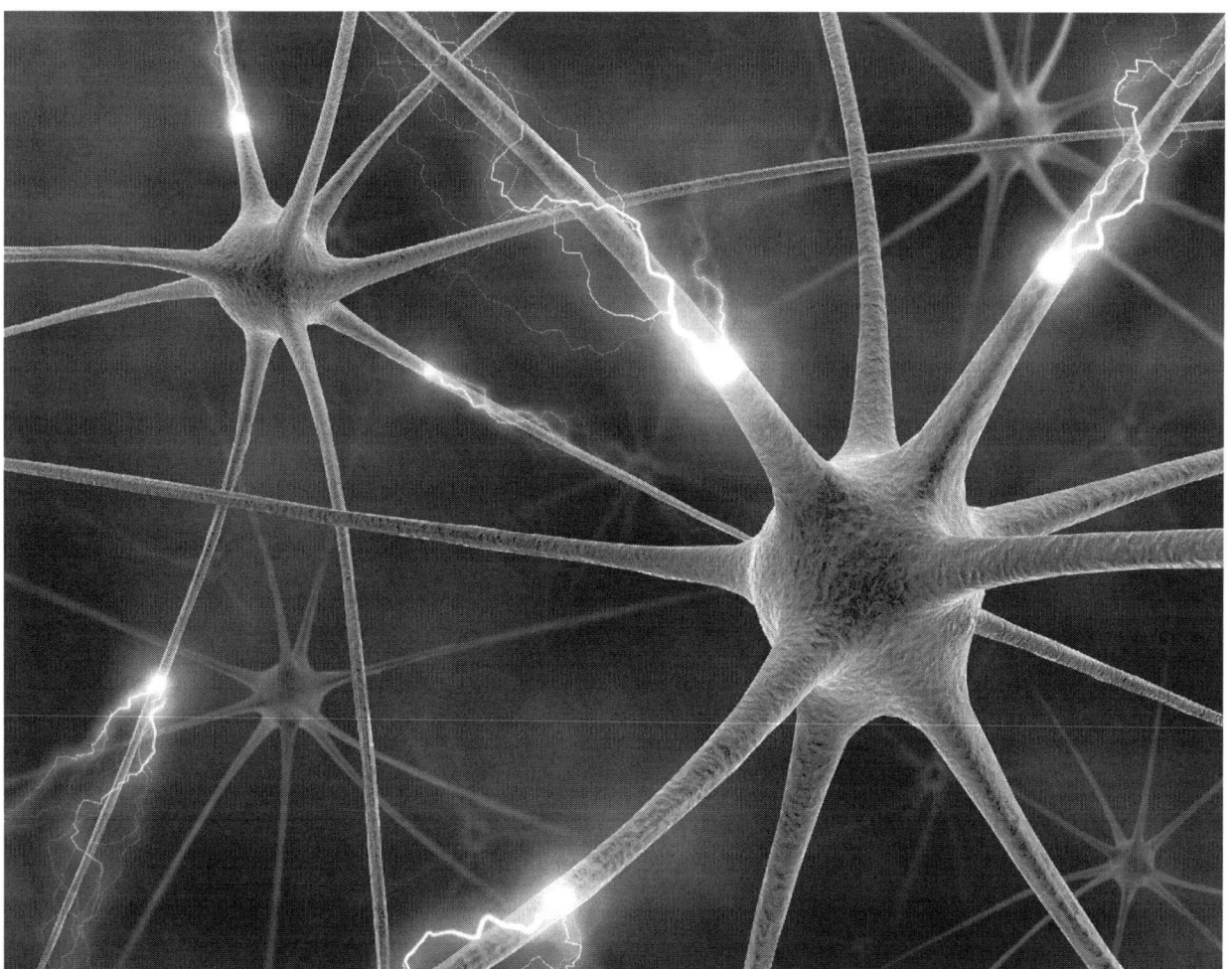

An illustration of hardwired neurons, transferring pulses and generating information. The electrical signal is transferred from one neuron's axon to another's dendrite across a gap known as a **synapse**. (iStockphoto)

interested in understanding normal brain processing as well as diseases. At each stage, mathematics plays a crucial role.

Brain Imaging Technologies

Magnetic resonance imaging (MRI) is commonly used today for noninvasive imaging of the internal structure of the human body; for example, to help determine if knee surgery or back surgery is warranted. In standard MRIs, a powerful magnetic field changes rapidly, and, by doing so, it manipulates the minute magnetic fields produced by protons in water molecules inside the body—a weak signal can be detected externally from the protons being flipped around by the strong magnetic fields. From these weak and indirect measurements, solving the inverse problem using mathematics related to calculus is used to create what appear as two-dimensional slices through the body. In the case of brain studies, the resolution of standard MRIs is adequate to see tumors but is too crude to see individual neurons or even to effectively track bundles of neurons. Since myelin is a fatty substance, water outside neurons will generally not diffuse into axons; rather, this water will tend to diffuse along the length of axons—the water percolates along axons or white matter.

Diffusion tensor imaging (DTI) uses a variation of a standard MRI to determine the diffusion direction, and hence determine bulk nerve fibers. While DTI can produce high-resolution images of nerve fibers, difficulties arise when fibers cross. Water diffusion in this case can now take multiple paths at the crossing points, and it is thus difficult to track nerve fibers at these crossing points. Diffusion spectrum imaging (DSI) involves more mathematics that determines more precisely how water diffuses and is not limited to thinking that water diffuses in only one direction. Roughly speaking, the mathematics is a mixture of calculus and statistical ideas, and it is interesting that two-dimensional ellipses and three-dimensional ellipsoids play a role in the mathematics of DTIs and DSIs. The resulting images of nerve fibers are visually striking.

Not all techniques for imaging neurons rely on such indirect approaches as conventional MRI, or the MRI-based DSI and DTI. Recall that those techniques are used primarily for imaging nerve fibers, not individual neurons. Techniques for higher-resolution imaging of actual neurons are somewhat direct. Jeff Lichtman and others developed the use of genes encoding three proteins that fluoresce in, essentially, the colors red, blue, and green. Genetically modifying mice with these genes, as well as an enzyme that randomly arranges the genes amongst neurons, allows for mice neurons to appear in one of now approximately 150 colors, creating what Lichtman has termed a "brainbow." Another approach relies on a genetically modified version of the rabies virus, which ordinarily is well suited for traveling from neuron to neuron on its journey to the brain. By tagging the modified virus with a fluorescent molecule, one obtains bright images of neurons connecting to just one other neuron to further aid in understanding neural connectivity. All in all, exceptionally striking images are displayed in many places including on the Internet. While the imaging of individual neurons is often more direct and makes less use of mathematics than the MRI approaches to imaging nerve fibers, much mathematics is subsequently used in automating the process of tracking neurons and nerve fibers and, ultimately, the connections found might be described and analyzed by an area of mathematics known as "graph theory."

The approaches to imaging nerve fibers discussed above, such as DTI and DSI, rely on MRI instruments; that is, they rely on indirect methods involving minute variations in very small magnetic fields from protons that are assailed by powerful externally generated magnetic fields. They use indirect information coming from throughout the local environment inside the brain, and mathematics is crucial to inverting the recorded data to recover what is going on at a particular location in the brain. But not all imaging inside the brain focuses just on the connections between individual neurons or bundles of neurons. Other areas of interest include determining which parts of the brain are stimulated at which times by which activities. Here, other inversion processes are used to see what is happening in the brain, including functional MRI (fMRI), positron-emission tomography (PET), and electroencephalograms (EEG). Blood oxygen level dependence (BOLD) uses MRI technology that takes into account the very slight differences in the magnetic fields from water molecules in blood, depending on whether the blood is carrying oxygen. Hemoglobin bound to oxygen is diamagnetic, essentially repelled by a magnetic field. Hemoglobin without bound oxygen is paramagnetic, or attracted to a magnetic field. In either case, it affects the overall magnetic fields from the water molecules.

This effect leads to functional MRI (fMRI), which is used to examine oxygenated blood in the brain. The principle is that high neural activity is probably associated with increased blood flow.

PET, another imaging approach, uses an analog of glucose with a radioactive fluorine atom attached. When it decays, it produces a particle known as a positron that is quickly annihilated upon encountering an electron, and two photons stream out in opposite directions. Photomultiplier detectors essentially notice the two photons, and mathematics is used to invert this problem and determine where the annihilation occurred, which presumably is near where the brain was consuming the glucose-like food. A single-photon to PETs, Single Photon Emission Computed Tomography (SPECT), is also utilized.

As the final example of imaging approaches, EEGs focus on using electrical activity recorded on the scalp to see what voltages are created by bulk neurons extending over somewhat larger regions of the brain, as neurons synchronize their electrical signaling. EEGs thus have less spatial resolution than some of the other imaging approaches. Magnetoencephalography (MEG) is a magnetic analog of the EEG in that it is also a noninvasive procedure. Rather than using electrodes attached to a person's scalp for measurements, as in an EEG, very precise superconducting quantum interference devices (SQUIDs) detect weak magnetic fields directly arising from electrical brain activity. A mathematical inverse process makes the externally obtained magnetic data usable and converts it to internal electrical activity. MEG typically offers greater resolution, so it can localize the electrical activity more precisely, than EEGs.

Much mathematics is used to model the flow of electrical impulses in the brain. Wave phenomena in the brain arise in varied contexts, from the propagation of signals down a neuron, to collective behavior of many neurons resulting in rhythmic activity. More specifically, the Nernst equation, named for Walther Nernst, and its generalization, the Goldman equation, named for David Goldman, help relate ion concentrations to voltages. How those voltages change in a neuron as it is stimulated by other neurons is modeled by the Hodgkin–Huxley set of equations, which are a calculus-based set of differential equations that resulted in a Nobel Prize for Alan Hodgkin and Andrew Huxley. Next, an area of mathematics known as "dynamical systems" helps model how the firing of individual neurons can naturally become synchronized and produce wave behavior at different frequencies, which are ultimately recorded on EEGs. Normal rhythmic activity is important in activities such as sleeping, breathing, or walking. Abnormal rhythmic activity is manifested in various diseases; for example, forms of schizophrenia, Parkinson's disease, and epilepsy demonstrate deviations from what is considered typical rhythmical behavior.

Neurology of Learning Mathematic

How the brain learns mathematics is another area of interest to researchers. Psychology and other social sciences bring light to bear on this subject but so too does the study of various neural pathologies. As an example, dyscalculia, which has been called a form of "number blindness" (by analogy to color blindness), is a pathology wherein individuals cannot acquire arithmetic skills. For instance, individuals fully capable of language communication who cannot tell if one whole number is larger than another or are unable to do 2-digit computations are considered "number blind." For other examples, there are cases of individuals with increasing difficulties with speech—primarily because of atrophy in the temporal lobes leading to dementia—having highly reduced vocabulary including an inability to name common objects, yet whose arithmetic abilities remained virtually flawless. Similarly, there are instances of autistic individuals essentially unable to speak or understand speech, who nevertheless can perform computations. Infants can notice when the number of objects in a display changes or when a number of objects are hidden behind a screen.

Finally, there are instances of stroke victims who have fully intact language but lack numerical skills, such as not being able to count past 4, or say how many days are in a week. These examples indicate that language is not crucial for arithmetic computations, and, further, language may not be necessary for learning to calculate. Generally speaking, computations seem to be localized to the parietal lobe at the top of the brain, whereas key language areas, such as Broca's (frontal lobe), named for Paul Broca, and Wernicke's (temporal lobe), named for Carl Wernicke, reside elsewhere. However, there are ongoing debates among neuroscientists regarding what the highlighted areas on images mean with regard to brain functionality.

A related issue is how mathematical thinking beyond the level of simple arithmetic evolved in humans, including its relationship with the development of language and increasingly abstract reasoning. There are different and intriguing hypotheses regarding why language evolved roughly 200,000 years ago, whereas various forms of numerical and algorithmic abstraction evolved within the past few thousand years.

Further Reading

Bookstein, Fred. "Morphometrics." *Math Horizons* 3 (February 1996).

Joint Policy Board for Mathematics. "Mathematics Awareness Month, April 2007: Mathematics and the Brain." http://www.mathaware.org/mam/07/announcement.html.

Martindale, Diane. "Road Map for the Mind: Old Mathematical Theorems Unfold the Human Brain." *Scientific American* 285, no. 2 (August 2001).

Schoonover, Carl. *Portraits of the Mind: Visualizing the Brain from Antiquity to the 21st Century*. New York: Abrams, 2010.

Sousa, David. *How the Brain Learns Mathematics*. Thousand Oaks, CA: Corwin Press, 2008.

Rick Kreminski
Dedicated to Wanda Kreminski (1925–2010)

Carbon Dating

Category: Weather, Nature, and Environment.
Fields of Study: Algebra; Data Analysis and Probability.
Summary: Exponential and logarithmic functions are used in carbon dating—a method of determining the age of plant and animal fossils.

As is demonstrated throughout this encyclopedia, mathematics provides explanations for many interesting physical phenomena, and enables humankind to better understand its surrounding world. One of our ongoing intellectual projects is simply to make sense of the world we inhabit, based on the evidence that surrounds us. As anthropologists, archaeologists, and geologists have worked to determine the age of the earth and to track the evolution of species, radioactive isotopes have played a prominent role in efforts to create a timeline that charts a wide range of historical developments. In particular, carbon-14 dating has provided a fundamental test enabling scientists to accurately date certain plant and animal fossils that are approximately 60,000 years old or less. Willard Libby was one of the first to research radiocarbon dating, and he won a Nobel Prize in chemistry. Carbon dating is not an exact science, and statistical methods are used to enhance the reliability of the methods.

The Mathematics of Carbon Dating

Left alone, a radioactive quantity will decay at a rate proportionate to the amount of the quantity present at a given time. More specifically, a radioactive chemical element (such as uranium) is one that is unstable; as it decays, it emits energy and its fundamental makeup changes as the mass of the element is changed to an element of a different type. Because such an element is losing mass at a rate proportionate to the available mass at time t, an exponential function may be used to model the amount of the isotope that is present.

Letting $M(t)$ represent the mass of the element at time t, it turns out that $M(t) = M_0 e^{-kt}$, where M_0 is the mass at initial measurement (at time $t = 0$), and k is a constant that is connected to the rate at which the element decays. Furthermore, k is tied to the isotope's half-life (the amount of time it takes for 50% of the mass present to decay). In the given model, if h represents the half-life, then when $t = h$, it follows that

$$M(h) = \frac{M_0}{2}.$$

That is, the equation

$$\frac{M_0}{2} = M_0 e^{-kh}$$

must hold. Dividing both sides by M_0, yields

$$\frac{1}{2} = e^{-kh}$$

and using the natural logarithm function, one may solve for k and thus rewrite the most recent equation as

$$-kh = \ln\left(\frac{1}{2}\right).$$

This can be rewritten as

$$k = \frac{-\ln\left(\frac{1}{2}\right)}{h}.$$

A property of the natural logarithm is that

$$-\ln\frac{1}{2} = \ln(2)$$

so that in slightly simpler terms,

$$k = \frac{\ln(2)}{h}.$$

Therefore, the model for radioactive decay of an element having half-life h is

$$M(t) = M_0 e^{-(\ln(2)/h)t}.$$

With this background in place, one is now ready to understand how carbon dating works.

All living things contain carbon, and the preponderance of the carbon present in plants and animals is its stable isotope, carbon-12. At the same time, every living being takes in radioactive carbon-14, and this carbon-14 becomes part of our organic makeup. While carbon-14 is constantly decaying simply by doing the normal things that come with being alive, each living organism continuously replenishes its supply of carbon-14 in such a way that the ratio of carbon-12 to carbon-14 in its body is constant.

When no longer living, a plant or animal lacks the ability to ingest carbon-14, and thus the ratio of carbon-12 to carbon-14 starts to change, and this ratio changes at the rate that carbon-14 decays. Chemists have long known that carbon-14 has a half-life of approximately $h = 5700$ years, and this knowledge, together with the exponential model

$$M(t) = M_0 e^{-(\ln(2)/h)t}$$

enables people to determine the age of certain fossils. Consider, for example, the situation where a bone is found that contains 40% of the carbon-14 it would be expected to have in a living animal. With less than half the original amount present, but more than 25%, it can be determined that the bone is somewhere between one and two half-lives old; that is, the animal lived between 5700 and 11,400 years ago.

Through our understanding of exponential functions and logarithms, this estimate can be made much more precise.

Specifically, let $t = 0$ be the year the animal died. The present year t satisfies the equation $M(t) = 0.4M_0$, since 40% of the initial amount of carbon-14 remains. From the model, it is known that t must be the solution to the equation

$$0.4\,M_0 = M_0 e^{-(\ln(2)/5700)t}.$$

First, divide both sides by M_0 to get $0.4 = e^{-(\ln(2)/5700)t}$ and then, taking the natural logarithm of both sides of the equation, it follows that

$$\ln(0.4) = \frac{-\ln(2)}{5700}t.$$

Thus, solving for t yields

$$t = \frac{-5700\ln(0.4)}{\ln(2)} \approx 7500 \approx \text{years}$$

and the skeletal remains have been dated according to their carbon content.

Limitations of Carbon Dating

Carbon dating does have some reasonable limitations. One of these involves the complications of measuring only trace amounts of carbon-14, and emphasizes the behavior of functions that model exponential decay. For each half-life that passes, half of the most recent quantity of the element remains. That is, after one half-life,

$$\frac{M_0}{2}$$ remains; after two, half of that amount,

or $$\frac{M_0}{4}$$ is left;

after three, $$\frac{M_0}{8}$$ is present.

The quantity rapidly diminishes from there. For instance, after 10 half-lives have elapsed, there is

$$\frac{M_0}{2^{10}} = \frac{M_0}{1024} \text{ or approximately}$$

$0.0009766 M_0$ left. Because each living organism only contains trace amounts of carbon-14 to begin with (of all carbon atoms, only about one-trillionth are carbon-14), after 10 half-lives elapse, the remaining amount of carbon-14 is so small that it is not only difficult to measure accurately, but it is difficult to ensure that the measured carbon-14 actually remains from the organism of interest and was not somehow contributed from another source. Ten half-lives is approximately 60,000 years, so any organism deemed older than that needs to be dated in another manner, typically using other radioactive isotopes that have considerably longer half-lives.

Finally, because radiocarbon dating depends on naturally occurring radioactive decay, its accuracy depends on such decay not being accelerated by unnatural causes. In the 1940s, the Manhattan Project resulted in humankind's development of synthetic nuclear energy and weapons; subsequent nuclear testing and accidents have released radiation into the atmosphere that makes the accuracy of carbon-14 dating more suspect for organisms that die after 1940.

New Developments

The exponential model $M(t) = M_0 e^{-kt}$ of radioactive isotope decay has enabled humans to better understand our surrounding world, and to know with confidence key information about the history of the existence of plant and animal life on Earth. Even today, there are new developments in the science of radiocarbon dating as experts work to understand how subtle changes in Earth's magnetic field and solar activity affect the amounts of carbon-14 present in the atmosphere. In addition to continuing to help analyze fossil histories, carbon-14 dating may prove an important tool in ongoing research in climate change.

The Accelerator Mass Spectrometry method of dating directly measures the number of carbon atoms rather than their radioactivity, which allows for the dating of small samples. Other methods under development include nondestructive carbon dating, which eliminates the need for samples. A group of Russian mathematicians have proposed a new chronology of history based on other methods for dating; however, many have dismissed their work as pseudoscience. Physicist Claus Rolfs explores methods to accelerate radioactive decay in the hope of reducing the amount of radioactive material.

Further Reading

"Archaeological 'Time Machine' Greatly Improves Accuracy of Early Radiocarbon Dating." *Science News Daily* (February 11, 2010) http://www.sciencedaily.com/releases/2010/02/100211111549.htm.

Brain, Marshall. "How Carbon Dating Works." http://www.howstuffworks.com/carbon-14.htm.

Comap. *For All Practical Purposes: Mathematical Literacy in Today's World*. 7th ed. New York: W. H. Freeman, 2006.

Connally, E. et al. *Functions Modeling Change: A Preparation for Calculus*. Hoboken, NJ: Wiley, 2007.

Matt Boelkins

Carbon Footprint

Category: Weather, Nature, and Environment.
Fields of Study: Algebra; Data Analysis and Probability; Measurement; Representations.
Summary: A carbon footprint is a mathematical calculation of a person's or a community's total emission of greenhouse gases per year.

Carbon footprint is intended to be a measure of the ecological impact of people or events. It is a calculation of total emission of greenhouse gases, typically carbon dioxide, and is often stated in units of tons per year. There is no universal mathematical method or agreed-upon set of variables that are used to calculate carbon footprint, though scientists and mathematicians estimate carbon footprints for individuals, companies, and nations. Many calculators are available on the Internet that take into account factors like the number of miles a person drives or flies, whether or not he or she uses energy efficient light bulbs, whether he or she shops for food at local stores, and what sort of technology he or she uses for electrical power. Some variables are direct, such as the carbon dioxide released by a person driving a car, while others are indirect and focus on the entire life cycle of products, such as the fuel used to produce the vegetables that a person buys at the grocery store and disposal of packaging waste.

The notion of a carbon footprint is being considered in a wide range of areas, including the construction of low-impact homes, offices, and other buildings. The design must take into account not only the future impact of the building in terms of carbon emissions, but carbon-related production costs for the materials, labor, and energy used to build it. Mathematical modeling and optimization helps engineers and architects create efficient, useful, and sometimes even beautiful structures while reducing the overall carbon footprint. Mathematicians are also involved in the design of technology that is more energy efficient, as well as methods that allow individuals and businesses to convert to electronic documents and transactions rather than using paper. These methods include using improved communication technology, faster computer networks, improved methods for digital file sharing and online collaboration, and security protocols for digital signatures and financial transactions. Manufacturers are increasingly being urged and even required to examine their practices, since manufacturing processes produce both greenhouse gasses from factory smokestacks and waste heat. Mathematicians and scientists are working on ways to recycle much of this heat for power generation. One proposed device combines a loop heat pipe, which is a passive system for moving heat from a source to another system, often over long distances, with a Tesla turbine. Patented by scientist and inventor Nikola Tesla, a Tesla turbine is driven by the boundary layer effect rather than fluid passing over blades as in conventional turbines. It is sometimes called a Prandtl layer turbine after Ludwig Prandtl, a scientist who worked extensively in developing the mathematics of aerodynamics and is credited with identifying the boundary layer.

Carbon footprints are calculated to include travel, fuel production, transportation, and storage. In Canada, mobility is the highest contributor to the national carbon footprint. (Photos.com)

These are in turn related to the Navier–Stokes equations describing the motion of fluid substances, named for mathematicians Claude-Louis Navier and George Stokes. The Navier–Stokes equations are also of interest to pure mathematics, since many of their mathematical properties remain unproven at the beginning of the twenty-first century.

Carbon Footprints of People

A calculation of the carbon footprints of different aspects of people's lives, and then the aggregate for a year, is always an estimate. For example, different towns use different methods for generating electricity. Entering data for an electric bill allows for a rough estimate of the household's carbon footprint, but not exact numbers, which would depend on the electricity generating methods. Houses contribute to carbon footprints through their building costs, heating and cooling, water filtration, repair, and maintenance—all of which use products with carbon footprints.

Travel is another major contributor to peoples' carbon footprints. Daily commutes and longer trips with any motorized transportation contribute to carbon dioxide emissions. When computing carbon footprints, fuel production and storage costs have to be taken into consideration.

The food that people eat contributes to the carbon footprint if it is transported by motorized vehicles before being eaten. The movement of locavores (people who eat locally grown foods) aims to minimize the carbon footprint of food. Also, different farming practices may contribute more or less to the carbon footprint of food.

The objects people use contribute to their carbon footprints. Recycling and reusing reduces the need for landfills, waste processing, and waste removal, all of which have carbon footprints. There are individuals and communities who avoid waste entirely; several countries, such as Japan, have plans to mandate zero-waste practices within the next few decades.

Economy and Policy

There are two main strategies for addressing carbon footprints. The first strategy is to lower the carbon footprint by modifying individual behaviors, such as traveling by bike, eating locally, and recycling. The second strategy is to perform activities with negative carbon footprints, such as planting trees, to match carbon footprints of other activities.

Some companies incorporate activities that offset the carbon footprint of their main production into their business plans, either lowering their profit margins or passing the cost to their customers. There are economic laws and proposals that attempt to integrate carbon footprint considerations into the economy, usually through taxes on use of fuel, energy, or emissions. Carbon dioxide emissions, in economic terms, are a negative externality (a negative effect on a party not directly involved in the economic transaction). Money collected through carbon taxes is generally used to offset the cost to the environment.

Emissions trading is another mathematics-rich area of dealing with carbon footprints economically. Governments can sell emission permits to the highest-bidding companies, matching their carbon footprints, and capping the total emission permits sold. This method allows prices of permits to fluctuate with demand, in contrast with carbon taxes in which prices are fixed and the quantities of emissions can change. Economists model the resulting behaviors, and advise policymakers based on the models' outcomes.

Marginal Abatement Cost Curve

"Marginal cost" is an economic term that means the change of cost that happens when one more unit of product is made, or unit of service performed. For physical objects, the curve is often U-shaped. The first units produced are very costly because their cost production involves setting up the necessary infrastructure. As more units are produced, and the infrastructure is reused, the price goes down until the quantities of production reach such levels that the logistic difficulties drive the price per additional units higher again.

A marginal abatement curve shows the cost of reducing emissions by one more unit. These curves are usually graphed in percents. For example, such a curve can be a straight line, with the cost of eliminating the first few percent of emission being zero or even negative. This happens because it can be done by changing practices within existing economic infrastructures, such as cheap smart switches into the residential sector's lighting grids. Additional lowering of the carbon footprint, however, requires deeper and costlier changes to the way of life. For example, there are relatively high costs involved in switching to wind and solar power, or switching to the use of crop rotations that do not require high-carbon fertilizers.

Country by Country

The average carbon footprint of citizens varies by country. For example, in late 2000s, the average annual carbon footprint of a U.S. citizen was about 30 metric tons per year, and a Japanese citizen about 10 metric tons per year. However, these calculations are extremely complicated because of global trade. For example, many developed countries "export" or "outsource" their carbon emissions to developing countries. Products imported from developing countries account for anywhere from a tenth to a half of the carbon footprints of developed nations.

International calculations indicate a strong correlation between the average carbon footprint of a country's citizen and the average per capita consumption. The higher the consumption rates, the higher the average carbon footprint.

The categories used for calculation for countries are similar to those used for individuals and include construction, shelter, food, clothing, manufactured products, services, transportation, and trade. The ratios of these items to one another in the carbon footprints vary by country. For example, the greatest item in the U.S. carbon footprint is shelter (25%), with mobility being second (21%). In contrast, Canada's greatest item affecting carbon footprint is mobility (30%), and its second greatest is shared between shelter and service (18% each).

Further Reading

Berners-Lee, Mike. *How Bad Are Bananas? The Carbon Footprint of Everything*. Vancouver, BC: Greystone Books, 2011.

Goleman, Daniel. *Ecological Intelligence: How Knowing the Hidden Impacts of What We Buy Can Change Everything*. New York: Broadway Books, 2009.

Maria Droujkova

Caves and Caverns

Category: Weather, Nature, and Environment.
Fields of Study: Algebra; Geometry; Representations.
Summary: Several metrics are used to describe caves while mathematical measurements can detect them.

Caves are underground spaces large enough for a human to enter. The science of studying caves is called speleology and the practice of exploring caves is spelunking. Caves can be formed through a variety of ways, such as solutional caves (made by rocks dissolving in acids in water) or littoral caves (made by waves pounding cliffs). They are also categorized by the passage patterns, such as angular networks or ramiform caves. Mathematical techniques are used to model and understand the structures and ages of caves and caverns. For instance, the topology of the cave highlights the number of tunnels and how they are connected while the geometry shows accurate distances, curvatures, and steepnesses. Statistical methods as well as fractal concepts of self-similarity have been used to estimate the number of entranceless caves. Archaeology has revealed that caves are among the oldest known human habitations.

Some researchers analyze ancient cave paintings for mathematical, astronomical, or geographical interpretations. Mathematical objects and mathematicians have also been connected to caves and caverns. The Lebombo bone was discovered in a Swaziland cave in the 1970s. It dates to approximately 35,000 b.c.e and is thought to be the oldest known mathematical artifact. The bone holds 29 tally notches and it has been compared to calendar sticks that are still in use in Namibia. In France, numerous mathematicians trained at École des Mines including Henri Poincare, who was employed as a mine engineer and was eventually promoted to inspector general.

Visitors today can enter Pythagoras' cave in Samos, where he apparently lived and worked on mathematics. In *The Republic*, Plato imagines chained prisoners in a cave who can only see shadows of the movement behind them. Similar metaphors continue to be explored in order to explain higher dimensional realities and other concepts in mathematics, physics, and philosophy, including investigations of quantum caves.

Geophysical Detection of Caves

The mapping of hidden caves and smaller karst formations is done for scientific and recreational explorations, as well as to ensure the stability of constructions, such as houses and bridges. Geophysical detection methods use contrasts in a physical property, such as electric resistance or density, between different parts of the underground medium. To detect variations, scientists measure microscopic changes in gravity caused by empty spaces, or transmit electromagnetic waves into the ground and measure their reflections. Another method is to transmit an electric current and measure changes in ground resistance. Seismic tomography depends on collecting massive amounts of data from inducing stress through boring holes, but it can be very accurate. All these methods depend on mathematical models of changes in physical properties between different surfaces.

(Photos.com)

Cave Measurement and Records

There are several metrics used to measure caves, including total length of passages, depth from the highest entrance to the lowest point; total volume, or height, depth, length, area; and volume of individual passages, shafts, and rooms. The deepest cave is 2191 meters meters (7188 feet) deep and the greatest total length cave is 591 kilometers (67 miles) long. These numbers are updated as more parts of caves are explored and new caves are discovered.

All geophysical techniques require contrasts of some physical property (density, electrical resistivity, magnetic susceptibility, seismic velocity) between subsurface structures.

Cave Patterns

The geometry of a cave depends on many geological factors, such as the structures dominant in the rock and the sources of water for solution caves. Spongework caves consisting of large, connecting chambers formed in porous rocks. If the rock also fractures easily, large chambers will be interspersed with long passages formed by fracturing in a pattern called "ramiform" (branchlike). Nonporous rock that fractures will produce a distinct pattern called rectilinear branchwork, with straight passages at angles to one another. Lava tubes are round in cross-section, long, and relatively even; they are formed by a lava flow that develops a hard crust.

Cave Meteorology and the Geothermal Gradient

Heat in caves comes from water or air entering the cave, or from overlying and underlying rock. Overlying rock does not transmit the surface heat well. For example, a difference of 30 degrees Celsius between day and night on the surface translates into 0.5 degrees Celsius difference one meter (3.28 feet) deep into limestone. Seasonal fluctuations penetrate deeper but still become negligible at depths of 10 or so meters (32.8 feet).

In most parts of the world, the temperature increases by about 25 degrees Celsius for every kilometer of depth, because of the molten interior of Earth, the rate called "geothermal gradient." As one goes deeper into a cave that starts at a sea level, the temperature first drops because of insulation from the surface but then increases because of the geothermal gradient. In areas of high volcanic activity near the surface, caves can be very hot, or even contain molten lava. Some of the deepest caves in the world are cold, because their entrances are high in the mountains.

Further Reading
Curle, Rane. "Entranceless and Fractal Caves Revisited." In: Palmer, A. N., M. V. Palmer, and I. D. Sasowsky, eds. *Karst Modeling, Special Publication* 5, Charlottesville, VA: Karst Water Institute, 1999.

Maurin, K. "Plato's Cave Parable and the Development of Modern Mathematics." *Rendiconti del Seminario Matematico, Università e Politecnico di Torino* 40, no. 1 (1982).

O'Connor, J. J., and E. F. Robertson. "Mactutor History of Mathematics Archive: Poincaré—Inspector of Mines." http://www-history.mcs.st-and.ac.uk/HistTopics/Poincare_mines.html.

Palmer, Arthur. *Cave Geology*. Trenton, NJ: Cave Books, 2007.

Maria Droujkova

Chemotherapy

Category: Medicine and Health.
Fields of Study: Algebra; Measurement.
Summary: Mathematical modeling has improved chemotherapy protocols and saved patients' lives.

Chemotherapy is the use of chemical drugs to kill cancerous cells in the body. Although cancerous cells are the target of chemotherapy, traditional chemotherapies do not distinguish between "good" and "bad" cells. Hence, chemotherapy often results in side effects, such as hair loss and toxicity damage to body organs. Because of these chemotherapy side effects, chemotherapy protocols attempt to kill as much of the tumor as possible while incurring as little damage to the patient as can be managed. Thus, chemotherapy regimens are managed according to different variables, including how much drug is given in a treatment, how frequently treatments are given, and the total number of treatments given. Historically, chemotherapy protocols were designed only through experimental data from clinical trials and practice. However, such experiments can be costly or even pose ethical dilemmas. Chemotherapy variables are quantitative—each lends itself to a mathematical understanding and description that can be used to model and simulate treatment experiments, adding to the information gained in clinical settings.

Mathematics in Cancer Chemotherapy News

Mathematics is becoming an increasingly powerful tool in cancer chemotherapy treatments, especially in the dosing and management of chemotherapy protocols. For example, in 2004, Dr. Larry Norton received the American Society of Clinical Oncology's David A. Karnofsky Award, which is given for an outstanding contribution to progression in cancer treatment. Norton's award is notable because of his quantitative contribution to the field of chemotherapy dosing. The National Cancer Institute has a Center for Bioinformatics that addresses the issue of systematically studying the vast amounts of data associated with cancer growth and treatment response.

Cancer Geometry and Treatment

Cancer cells appear visibly different in shape and structure than normal healthy cells. This fact helps practitioners identify unhealthy cells. Quantitative measurements are associated with the geometry and the complexity of cancer cells. These measurements are related to fractal geometry. Tumor fractal dimensions reflect more complex structures generally because of the arrangement of blood vessels in the tumor. Abnormal blood vessel arrangements inhibit the tumor's uptake of therapeutic drugs. This understanding has led to the use of anti-angiogenic drugs that inhibit the production of new blood vessels and lower the measurement of the tumor's complexity. These drugs can now be used in concert with other cancer treatments in order to create a more effective cancer-fighting regimen.

Cancer Growth and Chemotherapy Treatment

Historically, it was believed that cancer cells grew in an exponential manner over the entire period of a tumor's growth. In exponential growth, the doubling time of a population is constant. This belief affected the way that chemotherapy was delivered, since chemotherapy works by attacking rapidly dividing cells. If a tumor's growth rate were constant, there would be no difference in how many cells were killed during any chemotherapy treatment, regardless of the size of the tumor. This is the "log-kill" model of tumor growth.

However, in the mid-twentieth century, it was experimentally discovered that many tumors exhibited a different kind of population growth: "Gompertzian growth," named for Benjamin Gompertz. When populations grow in a Gompertzian fashion, they grow very rapidly at first—when the population is small. As the population size increases, the growth rate of the population slows. Thus, many tumors would have a smaller doubling time when smaller, and a larger doubling time when larger. Because chemotherapy attacks the most rapidly dividing cells, smaller tumors would be more susceptible to chemotherapy treatments. Thus, if a tumor has been reduced in size by one chemotherapy treatment, it would be better to give a second chemotherapy treatment as soon as possible without costing the patient in terms of healthy cell function. This Norton–Simon hypothesis, named for Larry Normal and Richard Simon, has led to a change in the frequency of standard chemotherapy regimens—the time between treatments was decreased in order to take advantage of the more rapid growth rate in the smaller tumor that had resulted from the previous treatment. This change in treatment timing has increased the survival time of patients undergoing chemotherapy treatments.

Looking Ahead

Although the Norton–Simon hypothesis is a prominent example of how mathematics has helped improve cancer chemotherapy treatments, there are ongoing studies by mathematicians to further improve treatment of cancer. Using a field of mathematics known as optimal control, some mathematicians study how to make chemotherapy treatments as ideal as possible. Although practitioners can make (and have made) use of the Norton–Simon hypothesis, the increase of chemotherapy treatments for a patient, while better, is not necessarily best. Using optimal control theory on mathematical models of cancer and cancer treatment, researchers can investigate the best timing and dosing strategies for chemotherapy based on the variables mentioned above. This work may even lead to determining cancer treatment plans based on a particular individual or a particular kind of cancer in the future.

Further Reading

Dildine, James. "Cancer and Mathematics." http://mste.illinois.edu/dildine/cancer/cancer.html.

Laird, Anna. "Dynamics of Tumour Growth: Comparison of Growth Rates and Extrapolation of Growth Curve to One Cell." *British Journal of Cancer* 19, no. 2 (1965).

Martin, R., and K. Teo. *Optimal Control of Drug Administration in Cancer Chemotherapy*. Singapore: World Scientific Publishing, 1966.

National Cancer Institute (NCI). *NCI Cancer Bulletin* 3, no. 18 (2006).

Schmidt, Charles. "The Gompertzian View: Norton Honored for Role in Establishing Cancer Treatment Approach." *Journal of the National Cancer Institute* 96, no. 20 (2004).

Piccart-Gebhart, Martine. "Mathematics and Oncology: A Match for Life?" *Journal of Clinical Oncology* 21, no. 8 (2003).

Angela Gallegos

Climate Change

Category: Weather, Nature, and Environment.
Fields of Study: Algebra; Calculus; Data Analysis and Probability; Problem Solving; Representations.
Summary: Mathematicians and scientists use sophisticated models to track and predict global climate change.

The term "climate change" refers to the changing distribution of weather patterns. Climate is considered to be the average of 30 years of weather. In other words, climate is the distribution from which weather is drawn. Global warming refers to the change in climate in such a way that warmer weather is increasingly likely. In fact, it is not just the warming itself that is of concern but also the rate of change of the warming process since ecological systems typically cannot adapt to a rapidly changing climate. According to the *2007 Synthesis Report* by the Intergovernmental Panel on Climate Change (IPCC), "Warming of the climate system is unequivocal, as is now evident from observations of increases in global average air and ocean temperatures, widespread melting of snow and ice and rising global average sea level." The main cause of changing climate is the increasing atmospheric concentrations of greenhouse gases (carbon dioxide, methane, and nitrous oxide), which effectively act as a blanket over the atmosphere.

The IPCC report noted, "There is very high confidence that the net effect of human activities since 1750 has been one of warming." The evidence for the warming of the climate includes more than the measurement of global average temperatures, as physical evidence such as glacier melt also exists. Most predictions of global warming are based on data models, and mathematics is used extensively to measure and quantify atmospheric carbon dioxide and aerosols, which are believed to add to the problem. The National Oceanic and Atmospheric Administration (NOAA) and the U.S. National Aeronautics and Space Administration (NASA) are two large federal agencies that are involved in the collection, analysis, and dissemination of climate data. They employ a diverse range of mathematicians, statisticians, scientists, and others, and they have partnerships with many academic institutions, government agencies, and businesses around the world. Researchers do agree, however, that the current and future consequences of climate change disproportionately impact the world's poor.

Climate as a Distribution

In order to understand global warming, it is important to understand that the term refers to a distribution. It is easy to dismiss the notion of global warming on a cold winter's day, a mild summer day, or any day where the weather is cooler than expected. In fact, an unusually hot or cold event is not evidence for or against global warming. To aid in understanding climate as a distribution, consider a set of 20 cards numbered 1–20. One card will be drawn at a time with replacement. The value 10.5 is the average of these cards; a card selected below 10.5 will represent a below-average temperature for the day, and one selected above 10.5 will represent an above-average for the day. Further, the farther away the value of a card is from 10.5 will represent a larger deviation from average temperature. This example represents a stable climate. Some days are colder than average, and others are hotter than average. But, over time, roughly an equal number of colder days and hotter days occur. Moreover, if the value of the cards were unknown, basic statistical sampling ideas could be used to estimate the average.

To represent a changing climate, start with the same set of cards and consider values below or above 10.5 as a colder or hotter day. But this time, every time a card is drawn and replaced, the next higher card will be added to the set. For example, after the first card is drawn, a 21 will be added to the set, then a 22 will be added after the second draw, then a 23 after the third, and so on. At first, this change would be barely noticed if at all since the cards drawn will be roughly equal above and below 10.5. After some time, however, one would start to question the assumption that 10.5 is the average. In this case, if the values of the cards were unknown, basic statistical sampling techniques could not be used to estimate the average since the average is in fact changing. In this example, if 10.5 is taken as the average, then values below 10.5 still occur but are becoming less likely. In other words, record lows can still occur—and will still happen— even though climate is warming.

To complicate this example further, consider this same experiment being performed simultaneously by 2000 people to represent different locations around Earth. When the set of cards have values from 1 to 100, one individual would have only a 10% chance of drawing a card below 10.5, but it is expected that approximately 200 of the 2000 experiments will draw a card below 10.5. In other words, even though climate is warming, there will still be places that have colder than average days.

In terms of actual weather, consider Figure 1, which provides the average monthly temperature anomalies in degrees Celsius for December 2009 compared to the average from 1951 to 1980. The month of December was slightly colder for most of the United States. The overall average for the world was 0.60 degrees Celsius higher than the baseline years. One month, or even one

Figure 1. Global temperature anomalies for December 2009.

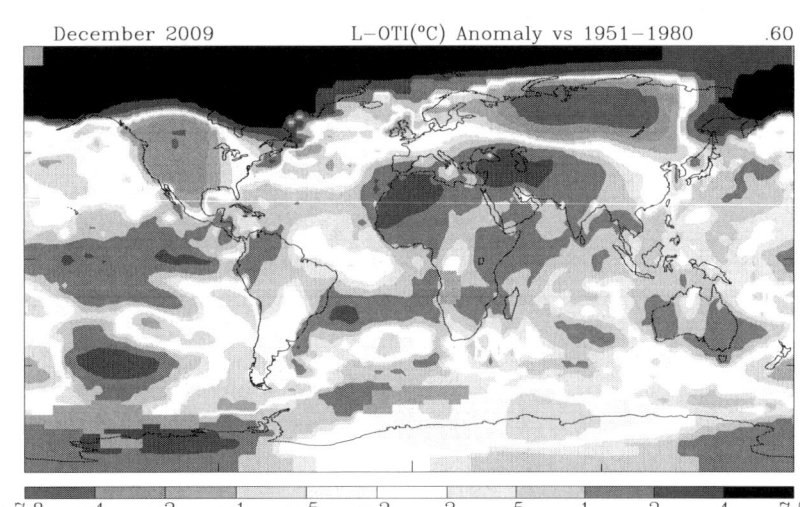

year or a few years, of above average temperatures does not provide conclusive evidence for or against global warming as these abnormalities could be explained as normal variations in weather.

Evidence of Warming

Calculating global mean temperatures each year provides one form of evidence for global warming. For example, Figure 2 displays mean global temperature anomalies dating to 1850. Even though the overall trend is upward, variation from one year to the next can go in either direction. Gerald Meehl, who has a Ph.D. in climate dynamics and works at the National Center for Atmospheric Research (NCAR), collected information from 1800 weather stations across the United States that have been operating since 1950. He and his colleagues looked at the ratio of record highs to record lows and grouped the ratios by decade. From the 1950s through the 2000s, the ratio of record highs to lows was 1.09:1, 0.77:1, 0.78:1, 1.14:1, 1.36:1, 2.04:1. From the 1950s through the 1980s, the ratios might be considered to be in the range of normal variation for a stable climate. On the other hand, by the 2000s, it certainly appears that the observations no longer represent normal variation, and that the climate distribution is getting warmer.

In Figure 2, the baseline is the average from 1961 to 1990, with regression lines for different time periods from 1850 to 2009, 1910 to 2009, 1960 to 2009, and 1985 to 2009. The data are the HadCRUT3 data set provided by the University of East Anglia Climatic Research Unit.

Beyond data, a warming climate should present physical evidence in the form of melting ice. Figure 3 is one of a number of glacier image pairs, which are pictures of glaciers taken from the same vantage but 40–100 years apart. The change is striking. Where there was once ice, there is now ocean water with the glacier retreating about seven miles. In the foreground, thick vegetation exists where there was once rock. This change is because of microclimate changes since the ice is no longer cooling that area. Along with melting glaciers, Arctic sea ice is decreasing rapidly and permafrost is melting. In fact, the entire village of Newtok, Alaska, must be relocated because the loss of permafrost has allowed the banks of the Ninglick River to erode.

Melting ice is just one source of evidence of a changing climate. During most of the twentieth century, sea level was rising at a rate of 0.07 inches per year, but by the 1990s that rate increased to 0.12 inches per year. In 2006, the National Arbor Day Foundation updated its plant hardiness zone maps, and most of the zones shifted northward. In other words, many plants can now be grown where they could not before because of their cold hardiness. There have already been observed shifts in species ranges, a northward shift, as well as shifts in phenology (seasonal biological timing) toward events such as early blooming. In fact, many species have seasonal behavior that is occurring 15–20 days earlier than the behavior occurred in the mid-twentieth century.

The general trend of warming is only part of the story. If the planet warmed a degree or two over millions of years, then ecological processes could adapt and societies could migrate. Figure 2 has least squares regression lines calculated over the time periods of 1850–2009, 1910–2009, 1960–2009, and 1985–2009. The four regression lines are as follows:

$$y_{159}(t) = 0.0041t - 8.67281$$

$$y_{100}(t) = 0.00750t - 14.75315$$

$$y_{50}(t) = 0.01364t - 26.96187$$

$$y_{25}(t) = 0.01801t - 34.67615$$

In each case, the slope of the line, with units of degrees Celsius per year, is increasing as the time periods are shortened toward more recent years.

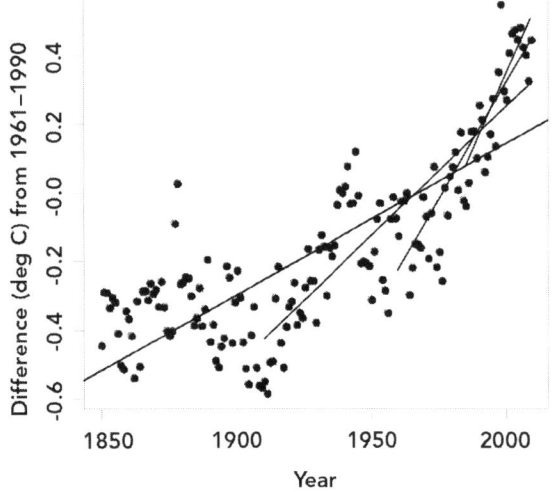

Figure 2. Global mean temperature anomalies.

More importantly, the 95% confidence intervals for the slopes are (0.00387, 0.00495), (0.00657, 0.00844), (0.01151, 0.01577), (0.01254, 0.02347), respectively. The first three intervals do not overlap, and so the slopes of the lines are significantly different. This provides evidence not only for overall warming but also that the rate of warming is increasing. Some species, trees for example, will simply not be able to adjust their ranges quickly enough to adapt to the warming climate.

Figure 3. Images of the Muir glacier taken from the same vantage on August 13, 1941 (left), and August 31, 2004 (right). (National Snow and Ice Data)

Climate Science

Climate models, which incorporate mathematical topics such as dynamical systems, statistics, differential equations, and applied probability, are used to predict future global average temperature.

Mathematician Ka-Kit Tung, in his book *Topics in Mathematical Modeling*, provides a simple climate model. The model is

$$R\frac{\partial}{\partial t}T = Qs(y)(1-\alpha(y)) - I(y) + D(y).$$

The left-hand side represents change in temperature. There are three basic terms on the right-hand side that contribute to temperature change. The first term has incoming solar radiation at the top of Earth's atmosphere,

$$Qs(y),$$

where the $s(y)$ term distributes the radiation differently depending on the latitude $y = \sin\theta$ with θ representing latitude. The term also takes into consideration how much radiation is absorbed

$$(1-\alpha(y))$$

where $\alpha(y)$ is the fraction reflected or albedo. The next term,

$$I(y)$$

represents outward radiation, and the last term,

$$D(y)$$

represents heat transportation from warmer latitudes to colder latitudes. In Tung's textbook, this simplified model is analyzed to gain understanding of possible locations in ice lines.

The more complex computer simulations that model climate are built with assumptions related to population growth and societal choices, such as energy use or technological change. These assumptions are then used to predict how greenhouse gases will increase. The effect of increased greenhouse gases in trapping heat is well understood, and in terms of the simple climate model above, the increase in greenhouse gases decreases outward radiation. Beyond that, the increase in carbon levels itself is a problem as oceans work to absorb some of this carbon in the form of carbonic acid. The increase in carbonic acid in the oceans increases the acidity levels, which damages coral, crustaceans, sea urchins, and mollusks.

For each scenario, many different models are considered, and the predictions are averaged to produce the graph on the left side of Figure 4. The three higher curves illustrate the average warming. On the right side of the graph is a range based on the various models. A distribution has been created, and based on the graph, one could say that, by 2100, global mean temperatures will increase between approximately 1.5 degrees Celsius and 3.5 degrees Celsius, but the distribution around the three scenarios presented is from approximately 1 degree Celsius to 6 degrees Celsius. The right side of Figure 4 presents the predicted temperature changes as a distribution across Earth, and it is predicted that the Arctic region will warm more than the equatorial region.

A key complication in climate modeling is the existence of feedback loops. A feedback loop is created when a change in one factor causes a change in a second factor that then either reinforces or diminishes the change in the first factor. While each scenario sets out greenhouse gas levels, the models must then attempt to take into account how warming may, in fact, increase warming or decrease warming. For example, one positive feedback loop involves melting ice. As ice melts, the Earth's albedo (reflectivity) changes so that less solar radiation is reflected out to space. In the climate model above, the $\alpha(y)$ term is decreased so that more solar radiation is absorbed. In other words, as the planet warms, ice melts. However, there are now fewer reflective white surfaces and more dark surfaces, which will then absorb even more solar radiation and increase the planet's warming. Another potential positive feedback loop arises from melting permafrost. As the permafrost melts, partially decomposed organic matter will decompose more fully and release carbon into the atmosphere. Even more uncertainty arises with the effect of clouds. Low clouds tend to cool by reflecting more energy than they trap, while the reverse is true for high clouds. As surface temperature increases, there is increased evaporation from the oceans, creating more water vapor and hence clouds. But the type of clouds that arise will depend on whether this is a positive or negative feedback loop.

Of course, to many people, an increase of a few degrees Celsius does not seem to be drastic enough to impact life on Earth significantly. But consider that during the twentieth century, global average temperatures increased by less than 1 degree Celsius. Nevertheless, there has already been observed disappearing glaciers, loss of Arctic sea ice, changing species habitat and phenology, and a new plant hardiness map. In fact, a difference of approximately 0.2 degrees Celsius was the difference between the Medieval Warm Period (c. 950–1250) and the cooling period (c. 1400–1700). The warm period led to the Norse migrating to Greenland and bountiful harvests and population increases in Europe. This period was followed by a cooling period that led to the collapse of the Norse Greenland society and starvation in Europe.

Impacts of Climate Change

The general consensus in the scientific community in 2010 is that warming has occurred and will continue to take place even with changes. Debate continues on precisely how much warming will occur and the exact nature of the ramifications. The questions are by how much, and what should people expect to happen? Species ranges are already changing, and, in some cases, species ranges are disappearing as appears to be the case for polar bears. Unfortunately, the speed of warming will lead to some species not being able to change their range quickly enough, resulting in extinction. The changing phenology is already causing ecological disruption. Some plants are blooming earlier, but the spe-

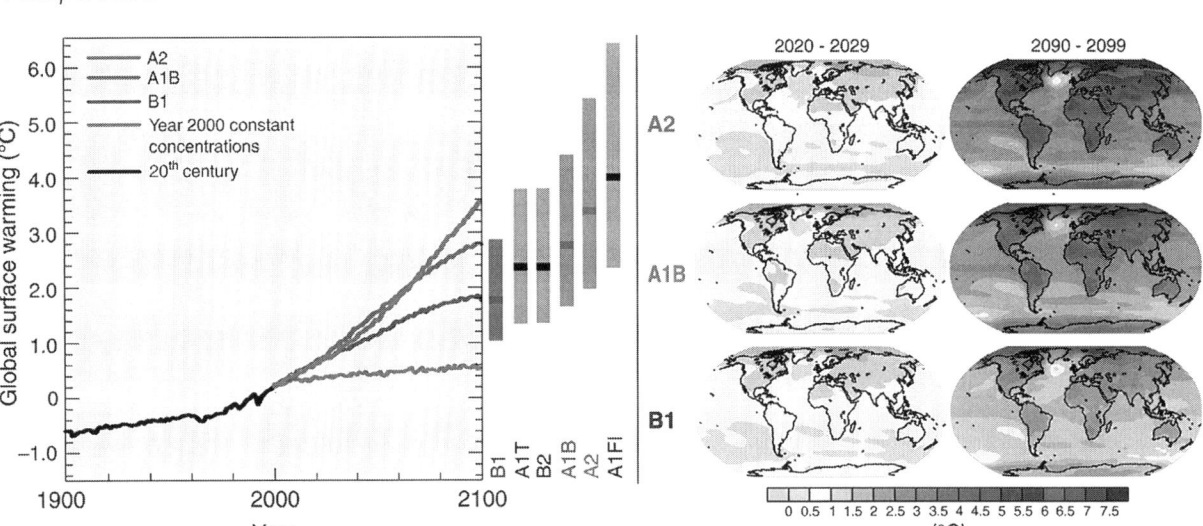

Figure 4. Climate model predictions of future average global temperature and distributional changes of temperature.

cies that feed on them are not arriving earlier, leading to decreased food supply.

As Figure 3 shows, ice is melting and more of that is expected. The loss of Arctic ice will decrease polar bear populations. The melting glaciers of the Tibetan Plateau are of particular importance. These glaciers are responsible for supplying water to about 2 billion people, and data suggests that the Tibetan Plateau is warming twice as fast as the global average. Once these glaciers are gone, so is the water supply. The melting of glaciers and land ice, along with the thermal expansion of water, will raise sea levels. One example is Bangladesh, which faces severe threats from sea level increases since millions of people live along a coastline that may be underwater in the future.

There are additional predictions as of 2010, based on models and scientific expertise. An increase of 2 degrees Celsius from pre-industrial levels would lead to a fall in agricultural yields in the developed world, a 97% loss of coral reefs, and 16% of global ecosystems transformed. With an increase of 3 degrees Celsius, few ecosystems could adapt and an additional 25–40 million people would be displaced from the coasts because of sea level rise. If global average temperatures rose to 4 degrees Celsius above pre-industrial levels, entire regions would be out of agricultural production, including Australia.

Climate Change and Societies

The joint science academies' statement on sustainability, energy efficiency, and climate protection issued in 2007 by the G8 nations and Brazil, China, India, Mexico, and South Africa, said that, "Many of the world's poorest people, who lack the resources to respond to the impacts of climate change, are likely to suffer the most." The warming of the planet will have some advantages and disadvantages, although there will be more disadvantages. Some warmer climate species will have expanded ranges and be able to thrive, while arctic species may lose their entire ecosystem. Some countries will be impacted more than others, and the wealthier countries will have a better ability to adapt. The examples that have been given here of societies that already have been or will likely be impacted are all examples of poorer societies.

The people of Newtok, Alaska, are poor; in 2010, Bangladesh ranked 183 in the world in terms of GDP per capita; and there is considerable poverty in regions in and around the Tibetan Plateau. Part of the tragedy is that these are not the people who are largely responsible for increasing greenhouse gases. China and the United States are the largest emitters of carbon dioxide, but on a per capita basis, the United States far exceeds China. In general, it is the industrialized nations that contribute the most to greenhouse gases. Figure 4 provides different models for future climate change, and these are primarily based on the models that predict future greenhouse gas emissions, and it is the more industrialized nations that have the resources to make reductions in these emissions.

Further Reading

Fleming, James. *Fixing the Sky: The Checkered History of Weather and Climate Control*. New York: Columbia University Press, 2010.

Intergovernmental Panel on Climate Change. "Climate Change 2007." http://www.ipcc.ch/pdf/assessment-report/ar4/syr/ar4_syr_spm.pdf.

Mackenzie, Dana. "Society for Industrial and Applied Mathematics: Mathematicians Confront Climate Change." http://www.siam.org/news/news.php?id=1131.

The National Academies. "Ecological Impacts of Climate Change." http://dels-old.nas.edu/climatechange/ecological-impacts.shtml.

The National Academies. "Understanding and Responding to Climate Change, 2008 Edition." http://dels-old.nas.edu/dels/rpt_briefs/climate_change_2008_final.pdf.

National Oceanic and Atmospheric Administration. "Climatic Data Center." http://www.ncdc.noaa.gov/oa/climate/globalwarming.html.

Neelin, David. *Climate Change and Climate Modeling*. New York: Cambridge University Press, 2011.

Oreskes, Naomi, and Erik Conway. *Merchants of Doubt: How a Handful of Scientists Obscured the Truth on Issues From Tobacco Smoke to Global Warming*. New York: Bloomsbury Press, 2010.

U.S. Global Change Research Program. "Global Climate Change Impacts in the United States." http://downloads.globalchange.gov/usimpacts/pdfs/climate-impacts-report.pdf.

Weart, Spencer R. *The Discovery of Global Warming: Revised and Expanded Editions*. Cambridge, MA: Harvard University Press, 2008.

THOMAS J. PFAFF

Clouds

Category: Weather, Nature, and Environment.
Fields of Study: Data Analysis and Probability; Geometry.
Summary: The formation and behavior of clouds can be mathematically modeled and studied.

Mathematics has been called "the science of patterns." In clouds and the atmosphere, generally there is no end to the patterns that may be observed, quantified, and more clearly understood using mathematics. Mathematicians have long modeled the behavior and structure of clouds.

Applied mathematicians continue to develop ways to detect clouds and quantify motion, composition, density, top altitude, and the distance between clouds, among other characteristics. In 1999, the U.S. National Aeronautics and Space Administration (NASA) launched the Multi-Angle Imaging SpectroRadiometer to measure environmental and climate data from nine different angles, including cloud data.

The U.S. National Oceanic and Atmospheric Administration (NOAA) is one of the largest organizations specializing in the study of the environment. In 2010, a NOAA team led by physicist Graham Feingold reported its findings that clouds form synchronous patterns, meaning that individual clouds in a group respond to signals from other clouds, an effect also observed in chirping crickets or flashing fireflies. This research has implications for interpreting climate change data. There are also mathematical objects such as point clouds that are of interest in geometry, imaging, and efficient distribution mining. Fractal clouds are appreciated for their mathematical properties and their artistic qualities.

Water in the Air

Air is composed primarily of nitrogen (78%) and oxygen (21%). Argon comprises nearly 1%, leaving little room for the remaining gasses, including carbon dioxide, ozone, and neon. This recitation, however, is for dry air. Water vapor, the invisible gas from which clouds are constructed, can account for 0% to 4% of any given parcel of air. In order to form a cloud, water vapor must change phase to either liquid water droplets or ice crystals.

The Transformation of Water into Clouds

The amount of water vapor that can be held in a parcel of air is determined primarily by the temperature of the air; warm air can hold more and cold air less. The amount of water vapor held in a parcel of air is identified by the mixing ratio:

$$w = \frac{\text{grams of water vapor in a parcel}}{\text{kgs of dry air in the same parcel}}.$$

The amount of water vapor a parcel of air *can* hold is called the "saturation mixing ratio":

$$w_1 = \frac{\text{grams of water vapor in a saturated parcel}}{\text{kgs of dry air in the same parcel}}.$$

Relative humidity is a measure of how much vapor a parcel of air is holding compared to how much it could possibly hold and is expressed algebraically as

$$\text{RH} = 100\frac{w}{w_1}.$$

The dew point is the temperature at which a parcel of air becomes saturated. At this point, the saturation mixing ratio and the actual mixing ratio are equal to one another, and the relative humidity is therefore 100%. A further drop in temperature should produce condensation as water changes phase from vapor to liquid cloud droplets or solid ice crystals—a cloud is born.

The Unstable Atmosphere

Clouds are often the result of lifting in the atmosphere. When a parcel of air rises, it generally cools, and this cooling produces condensation. The way in which the lifting is accomplished can lead to dramatic differences in the appearance of the cloud. When whole layers of air are gently lifted in an atmosphere that is stable, stratus clouds are formed, whereas the more dramatic vertical structure of a cumulus cloud comes from runaway convection, a self-perpetuating process that can build clouds more than 12 kilometers (km) or 40,000 feet tall.

What is a stable atmosphere? Temperatures generally decrease with height. The rate of change is, of course, variable but it is referred to as the "lapse rate" (Γ) of the atmosphere. A parcel of air, distinct from the air that surrounds it, may be forced to rise or descend and will cool or warm as a result. Pressure generally decreases with height, and a parcel that rises into a zone of lower pressure will expand, doing work on the environment and therefore cooling. The rate at which a parcel of air cools as a result of this sort of ascension is known as the "dry adiabatic lapse rate" (Γ_d) which is approximately 10 degrees Celsius per km. When the dew point is reached in the parcel and condensation occurs, latent heat is released as a result of the phase change and the parcel is warmed.

The result is a lower lapse rate, the saturated adiabatic lapse rate (Γ_s). The saturated lapse rate depends on the amount of moisture being condensed but 6 degrees Celsius per km may be used as a rough estimate.

Now if $\Gamma < \Gamma_d$, the atmosphere is stable because unsaturated air that is made to rise will cool at approximately 10 degrees Celsius per km and will find itself in air that is increasingly warmer than itself. The greater the difference $\Gamma_d - \Gamma$, the greater the force restoring the parcel to its previous altitude. The force may be quantified as

$$\frac{g(\Gamma_d - \Gamma)\delta z}{T}$$

where g is the gravitational constant, T is temperature, and δz is a small upward displacement of the parcel from its equilibrium level. Consider the implications of a temperature inversion in which temperature actually increases with height and Γ is a negative quantity. Now consider a situation in which the atmosphere cools strongly with height, that is $\Gamma > \Gamma_d$. Then, the restoring force becomes negative. Air that rises becomes warmer than its surroundings and so continues to rise. This leads to the runaway convection that builds the towering cumulonimbus clouds that can produce thunderstorms, lightning, and hail.

Further Reading

Adam, John. *Mathematics in Nature: Modeling Patterns in the Natural World.* Princeton, NJ: Princeton University Press, 2003.

Feingold, Graham, et al. "Precipitation-Generated Oscillations in Open Cellular Cloud Fields." *Nature* 466, no. 12 (August 2010).

Wallace, John M., and Peter V. Hobbs. *Atmospheric Science.* Burlington, MA: Academic Press, 2006.

Mark Roddy

Coral Reefs

Category: Weather, Nature, and Environment.
Fields of Study: Connections; Data Analysis and Probability; Geometry.
Summary: Mathematics helps describe and explain the formation of coral reefs.

Coral reefs are complex stony structures made of exoskeletons of coral polyps. Colonies of polyps form corals, with their stony parts consisting of calcium carbonate. All polyps in a single coral are genetically identical. Polyps get their energy from photosynthesis of their internal symbionts, one-cell algae living in the polyps. Some corals also have stinging tentacles for catching plankton, and can be painful for people to touch. The development and growth of coral reefs and atolls was fiercely debated in the nineteenth and early twentieth centuries. Charles Darwin argued in his 1842 publication *Structure and Distribution of Coral Reefs*, based on his personal observations, that the geometry of coral reefs resulted from the natural geological subsidence of oceanic islands.

In other words, coral reefs formed around islands, growing as the islands sank away. Darwin's chief opponent in this debate was Alexander Agassiz, who advocated the theory that coral reefs were not wholly dependent on subsistence for their formation but rather arose from a variety of geological and biological factors. Agassiz collected data from nearly every coral reef on Earth before his death in 1910, but none of his research had been published at that time. Contemporaries of both Darwin and Agassiz were inhibited by the inability to collect data other than observations and relatively shallow rock samples. In the 1950s, geologist Harry Ladd conducted tests in conjunction with the U.S. War Department, including boring thousands of holes in the coral of Eniwetok Atoll. Ladd's drill went to a depth of nearly 5,000 feet before finally passing completely through the coral into the soil below, confirming in many scientists' minds that the atoll had been built up as the land had sunk away. Ladd purportedly erected a sign on Eniwetok that read, "Darwin was right!"

Measurements and Variables

The shape of a coral reef is determined by the sea floor and the historical changes in sea levels. Reef scientists recognize three main shape types: fringing reefs, barrier reefs, and atolls. Fringing reefs stay close to shores, and their shape is determined by the shore they circle. Barrier reefs start as fringing reefs, but as the water levels rise relative to the shore, there are deep, large lagoons separating the shore and the reef. When volcanic islands completely subside underwater, their fringing or barrier reefs can stay near the surface, forming a circular lagoon. Such reefs are called "atolls."

In most places, sea levels rise over the land. The speed of reef growth depends on multiple variables, including temperature, water salinity, water clarity necessary for photosynthesis, and wave action. Reefs can grow up to 25 centimeters (about 10 inches) per year in height. Reefs cannot grow faster than sea levels rise, because the polyps can survive out of water only for a short time—for example, during the low tide. When the speed of reef growth matches the rise of the sea level, they are called "keep-up reefs." When the speed of reef growth is slower than the rise of the sea level temporarily, reefs may become either "catch-up reefs" when the speeds eventually match, or "drowned out" reefs that die as they are submerged too deeply. Global warming threatens to increase the rate at which sea levels rise beyond the speed of reef growth.

Because reefs need clear waters for photosynthesis, they grow in the parts of the ocean that are relatively nutrient-poor. However, reefs themselves support rich and diverse ecosystems—the contradiction called "Darwin's paradox." Reefs underlie less than 1% of the world's ocean beds but host about 25% of the marine species. They are called "underwater rainforests" because of their active biomass production, measured in weight per area per day.

Coral reefs have high fractal dimensions; in other words, their surface is rough, wrinkled, and uneven. This characteristic explains why corals thrive in moving waters. The fractal-like coral surfaces break the still water barrier surrounding them, with any agitation of water creating and amplifying turbulence. This turbulence means more water moves through the polyps, delivering nutrients to them and removing sediments that could prevent photosynthesis.

Mathematical Models

Coral reefs are vulnerable to storms, tsunamis, and other strong natural events. By modeling reef damage, it is possible to intervene, and to preserve some reefs that would otherwise be destroyed. Existing models include equations that measure the forces applied to reefs, and the forces reefs can withstand.

The ratio between the area of attachment of a reef and its total surface area plays a role in the models. The higher the surface area of the reef, the higher the pressure storms apply to it. On the other hand, the higher the area of attachment, the more force it takes to detach the reef. By modifying these variables, as well as the force of the storm, oceanologists can predict what happens to particular reefs. Moreover, with more compu-

Reefs need clear waters for photosynthesis and can be modeled as interesting hyperbolic structures. (Photos.com)

tation power comes the opportunity to model detailed shapes of reefs, individual currents, and other local variables, making predictions more precise.

Dynamic systems of differential equations are the area of mathematics applicable to complex ecosystems such as coral reefs. More deterministic models such as algebraic or simple differential equations do not capture the reality as well.

Hyperbolic Crochet Coral Reef Project

The crocheted coral reef is a collaborative project with hundreds of contributors and several exhibits worldwide, and is coordinated by the Institute for Figuring. It demonstrates hyperbolic geometry, which is a non-Euclidian geometry discovered about 200 years ago and found in nature—including corals. "Hyperbolic crocheting," the process for modeling corals, was first described in the late 1990s. It involves a simple repeating algorithm with introduced "mutations" that produce varied forms.

The models explore mathematical entities that can be found in coral reefs, such as the hyperbolic radius of curvature, pseudospheres, hyperbolic planes, and geodesics.

Further Reading

Dobbs, David. *Reef Madness: Charles Darwin, Alexander Agassiz, and the Meaning of Coral.* New York: Pantheon Books, 2005.

Institute for Figuring. "Hyperbolic Crochet Coral Reef." http://crochetcoralreef.org.

Sale, Peter. *Coral Reef Fishes: Dynamics and Diversity in a Complex Ecosystem.* San Diego, CA: Academic Press, 2002.

Maria Droujkova

Crystallography

Category: Weather, Nature, and Environment.
Fields of Study: Geometry; Measurement; Number and Operations; Representations.
Summary: Various mathematical principles are inherent in the structure of crystals and are used to study and classify them.

Crystallography is the study of the periodic structural arrangements of particles in solids. The first discoveries of the crystallographic structure of materials were made in the early twentieth century with the X-ray diffraction technique pioneered by Max van Laue. Solids that have crystal structures have a sharp melting point, which distinguishes them from amorphous substances, such as glass, which has neither a sharp melting point nor a crystal structure.

All matter tends to crystallize, since a crystal form is the lowest energy state. In reality, most physical crystals will have flaws rather than a perfect geometric structure. The chemical composition of a substance does not determine its crystal form. Calcareous spar, for example, has at least three distinct crystal types. Although crystals exist in three dimensions, some substances, such as graphite, form strong bonds between molecules in a plane, and only weak bonds between parallel planes. Mathematics is inherently connected to crystallography, as mathematicians describe and classify crystal structures and also use crystallographic methods to solve mathematical questions, such a packing problems. Despite almost a century of the existence of the modern science of crystallography, scientists do not have a good understanding of how local ordering principles produce large-scale order.

Lattices

The first consideration in crystal structure is the lattice, also known as the Bravais lattice, after August Bravais. There are 14 types of lattices. In a crystal structure, a translation is a motion in space in a certain direction through some distance. The arrangement of atoms, ions, and molecules must be periodic, and there must be three nonunique axes of translation. An axis of translation specifies a direction in which the structure repeats. If the whole structure is moved the proper distance in the direction of an axis, it will exactly cover itself. The lattice can be considered to be all the points to which any given particle can be translated by a translation, which also moves the entire crystal structure onto itself. Thus, the lattice consists of all the points that a given point or particle is moved to by a translation. From every point in the lattice, the view of the rest of the crystal is exactly the same. The portion of the crystal obtained by starting with a particle and moving it the smallest possible distance in each of the three translation directions is known as the unit cell.

Symmetries in Crystals

The geometry of a crystal structure is characterized by its symmetries. Besides translations, other symmetries include reflections in a plane, rotations through an angle about an axis, glide reflections (translation combined with a reflection), and screw translations (translation with a rotation). A crystal structure can only have rotations that are one-half, one-third, one-fourth, or one-sixth of a complete revolution. Mathematically speaking, two crystallographic structures are the same if their symmetries are the same. A collection of symmetries for an object is called a "symmetry group." Yevgraf Federov and Arthur Schoenflies, in the late 1800s, independently discovered that there are 230 distinct crystallographic symmetry groups in three-dimensional space.

Other Crystals

Wilson Bentley provided a wealth of insight into the structure of snow crystals using a photographic microscope, taking thousands of photographs of individual snowflakes over the course of 50 years. His photographs show that although snowflakes always have a basic hexagonal symmetry, they exhibit an endless variety of detail and seem to have a limitless number of forms. The simpler snowflakes grow slowly at high altitudes in low temperatures, and the more complex ones form at higher temperatures at greater humidity. Besides direct examination, information about the structure of snowflakes has been deduced by the forms of halos that they cause around the sun and moon.

In recent years, substances such as various aluminum alloys have been discovered to have regularity of structure but no translational symmetry. These substances are called "quasicrystals," and unlike true crystals, they can have 5-fold, 8-fold, 10-fold, or 12-fold rotational symmetry.

Further Reading

Bentley, W. A., and W. J. Humphries. *Snow Crystals*. New York: Dover Publications, 1962.

Burke, John G. *Origins of the Science of Crystals*. Berkeley: University of California Press, 1966.

Engels, Peter. *Geometric Crystallography*. Dordrecht, Holland: D. Reidel, 1986.

Kock, Elke, and Werner Fischer. "Mathematical Crystallography." http://www.staff.uni-marburg.de/~fischerw/mathcryst.htm.

Lord, Eric A., Alan L. Mackay, and S. Ranganathan. *New Geometries for New Materials*. Cambridge, England: Cambridge University Press, 2006.

Steven R. Edwards

Deforestation

Category: Weather, Nature, and Environment.
Fields of Study: Algebra; Measurement; Problem Solving.
Summary: Mathematicians study and model many aspects of deforestation.

Deforestation is the removal of forests by logging or burning. While some deforestation can occur accidentally as a result of wildfires, most is deliberate. Trees may be sold for lumber or charcoal, and land may be cleared for housing, farming, or pasturing livestock. Trees may also be removed for beneficial purposes, such as directing water flow or controlling future forest fires. Many people believe that deforestation is a significant factor in climate change and biodiversity loss, and research has shown that deforested regions are much more vulnerable to soil erosion and desertification.

While logging is linked to deforestation in the popular imagination, the United Nations Framework Convention on Climate Change actually found that in the early twenty-first century, logging actually accounted for less than 20% of deliberate deforestation. In contrast, commercial agriculture claimed about one-third of deforested lands and subsistence farming nearly one-half. This statistic indicates one reason why deforestation is increasing primarily in relatively poorer countries. However, within an industrialized country, like the United States, logging and clearing land for housing or other real estate development account for far more deforestation than subsistence farming, which few Americans have practiced since the dawn of the twentieth century. Mathematicians study and model many aspects of deforestation, including possible causes and the biological, geological, social, and economic effects; uses of deforested land; patterns of regrowth and biodiversity in areas where the forest has been allowed to return; and spatial mapping and visualizations of geographical regions before, during, and

Many people believe that deforestation is a significant factor in climate change and biodiversity loss, and research has shown that deforested regions are much more vulnerable to soil erosion and desertification. (iStockphoto)

after deforestation. Data collection, statistical analyses, and spatial dependency analyses, as well as stochastic spatial modeling, linear programming, geometry, and digital image analysis, are all mathematical methods that have played a role in such analyses.

Environmental Effects

Deforestation is implicated in numerous environmental problems. The relationship between the forest and atmospheric carbon dioxide, for instance, is complicated. While they are alive and actively growing, trees remove carbon dioxide from the atmosphere, store it as carbon, and release oxygen back into the atmosphere through respiration. This process reduces the amount of greenhouse gases in the atmosphere, and this basic dichotomy—plants breathing in carbon dioxide and releasing oxygen, while humans and animals do the opposite—has long been taught to schoolchildren as the critically interdependent relationship between flora and fauna on Earth. In the early twenty-first century, the world's forests store roughly three-quarters or greater of aboveground and soil carbon. When trees are cut down and burned, they release their stored carbon back into the atmosphere. When trees die and decay, they do the same, as fungi and bacteria break down the carbon products into carbon dioxide and methane. Their effect on the world's oxygen supply is actually very minor—the amount of oxygen they release is not as significant as the amount of carbon dioxide involved in a tree's lifespan.

But cutting trees down and turning them into long-lived products (using them to build houses, for instance) stores the carbon just as efficiently. For forests to continue to take carbon dioxide in from the atmosphere, the trees must be harvested regularly—with new trees planted—so that there are always actively growing trees.

Left to their own devices, mature forests cycle through periods as carbon dioxide sources (when the carbon dioxide released by decaying or wildfire-burned trees exceeds that taken in by growing trees) and sinks (when the net carbon dioxide release is negative).

The greatest amount of carbon dioxide is taken in by deciduous trees when spring leaves are growing, which results in an observable dip in the Keeling Curve (a graph that tracks variation in the concentration of atmospheric carbon dioxide from 1958 onward). The dip is mirrored by a rise corresponding to the release of carbon dioxide back into the air every fall when these leaves fall and decay. The curve is named for Charles Keeling, a University of California, San Diego, oceanographer whose observations helped bring global attention to anthropogenic climate change. Measurements continue to be taken at Mauna Loa, in Hawaii, and those data have shown a roughly 20% to 25% increase in the amount of atmospheric carbon dioxide between 1958 and 2010. There have been no declining trends in that time, countering the pre-Keeley claim that an apparent rise in carbon dioxide atmospheric concentration was the result of random fluctuations. Periodic local decreases and increases of about 1% to 2% are associated with seasonal cycles.

Anti-Deforestation Efforts

Recent efforts to reduce greenhouse gas emissions, and international agreements binding countries to do so, have brought more focus to the task of accurately measuring those emissions. It came to light in 2010 that Australia's efforts to reduce emissions in order to comply with the Kyoto Protocol goals were hampered by their inaccurate measurement of deforestation emissions. Since 1990, Australia has had the highest rate of deforestation in the developed world, and thus is the only developed country targeting deforestation emissions as its primary way of reducing overall emissions. But its inability to generate an accurate figure of what those emissions currently are, to establish a baseline, or reliably measure them in the future, has thrown a wrench in its efforts.

Data Collection and Mathematical Modeling

The highly complex nature of forest ecosystems and even individual trees makes it virtually impossible to collect complete data on the system dynamics of natural forests. As a result, investigations of long-term dynamics rely heavily on scientific inference. One way of making any estimate, heavily relied on when considering the environmental costs of possible actions, is through ecosystem modeling, which constructs mathematical representations of ecosystems. The entire ecosystem need not be represented (though this leaves open the possibility of unforeseen consequences in parts of the ecosystem not modeled). Typically, models are constructed to examine the inventory of a specific chemical in the environment, like carbon, nitrogen, phosphorous, or a toxin. The ecosystem is reduced to a set of state variables that describe the state of a dynamic system, like the population of a specific species or the concentration level of a particular substance.

Mathematical functions define the relationships between those variables, such as the relationship between new leaf growth and carbon dioxide intake. A usable model typically requires many variables and much fine-tuning to affirm that the relationships have been defined accurately, and, in some cases, a model may be constructed simply to test a hypothesis about those relationships, by comparing the behavior of the model ecosystem to the real one. For example, mathematician and ecologist Nandi Leslie developed mathematical models using techniques such as spatial statistics, mean field and pair approximation, and the theory of interacting particle systems to investigate questions about forest fragmentation and degradation, ecology and biodiversity in lands reclaimed by forests, and landscape-level impact of land-use activities in Bolivia and Brazil. Leslie is included on a Web site called Mathematicians of the African Diaspora and is the daughter of mathematician Joshua Leslie, who has published widely in the fields of algebraic and differential geometry. The applications of modeling in deforestation are as broad as the types of models. Some mathematicians have used calculus to measure tree density, including the number of trees per acre and the quantity of foliage. Logistic functions have been used to estimate insect density or infestations. Many linear and nonlinear modeling techniques, like regression analysis, are widely employed to help reveal and explain associations between multiple variables, such as social choices and government policies; economic measures; environmental measures; geographic features, like altitude and slope; and human constructions, like roads. These models are then frequently used to forecast important quantities of interest, like deforestation rates and the overall proportion of deforested land. However, inappropriate extrapolations and generalizations can lead people to make inaccurate predictions or conclusions. For example, extrapolations from exponential models tend to lead to overestimation of future values. This has an impact on contentious and world-reaching scientific debates, such as global warming.

Further Reading

Babin, Didier. *Beyond Tropical Deforestation: From Tropical Deforestation to Forest Cover Dynamics and Forest Development*. Paris: UNESCO, 2005.

Fowler, Andrew. *Mathematical Models in the Applied Sciences*. New York: Cambridge University Press, 1997.

Harte, John. *Consider a Spherical Cow: A Course in Environmental Problem Solving*. Sausalito, CA: University Science Books, 1988.

Shugart, Hurman. *A Theory of Forest Dynamics: The Ecological Implications of Forest Succession Models*. Caldwell, NJ: Blackburn Press, 2003.

Bill Kte'pi

Diagnostic Testing

Category: Medicine and Health.
Fields of Study: Data Analysis and Probability; Measurement; Number and Operations.
Summary: Diagnostic tests rely on statistics from clinical research to predict the presence or severity of a disease in a specific patient.

The ability of humans to detect and treat diseases has advanced considerably in the past two centuries, with the discovery of underlying causes, such as microorganisms, and treatments, like antibiotics, as well as methods for diagnosing injury and disease. In medicine, a diagnostic test is in an instrument used to detect or predict the presence or absence of disease or the severity of disease.

The instrument used may take a variety of forms, including a patient inventory or a mechanical device. In clinical research, it is common practice to assess the quality of such instruments relative to established gold standards.

Here, the intention is often to replace a traditional method by a newer one that offers greater benefits to health providers or patients, including cost reduction and less physical or psychological discomfort.

It may be of interest to use the diagnostic tool to predict outcomes based on existing symptoms. In this case, the gold standard is used to confirm patient outcomes for comparison with test predictions based on surrogate measures.

Common measures of instrument quality include reliability, validity, sensitivity, specificity, positive predictive value (PPV), and negative predictive value (NPV). Strictly speaking, these measures apply specifically to the scores forthcoming from the instruments rather than the instruments themselves, as they are based on studies applied to a specific sample of patients. Mathematicians and statisticians are essential partners in creating many diagnostic tools, such as magnetic resonance imaging, as well as for developing and refining the measures that allow clinicians and researchers to determine the efficacy of diagnostic instruments. They also help design experiments in which new instruments are tested and compared.

Nursing and other healthcare education programs frequently require courses in mathematics or statistics, and the field of biostatistics is one of the fastest-growing occupations in the late twentieth and early twenty-first century.

Reliability represents the reproducibility of the test outcomes. A simple case involves estimation of the extent of chance-corrected agreement in the interpretation of categorical findings from medical images derived from patients. Here, agreement might be measured across different clinicians based on a single imaging procedure or alternatively, across different imaging procedures. In such cases, an appropriate choice of Kappa statistic or intra-class correlation coefficient may prove helpful. For continuous data, the Bland–Altman method has also proved particularly popular in measuring agreement across different methods. This is especially so within medicine, where for example, there may be a need to compare residual tumor sizes obtained using magnetic resonance imaging, and pathologic findings (the gold standard) in breast cancer patients who have undergone neoadjuvant (preoperative) chemotherapy.

The remaining measures above represent the accuracy of the test outcomes. Validity, which is a function of reliability, represents the extent to which the diagnostic test measures what is intended and is particularly relevant in psychological testing. Sensitivity (specificity) measures the proportion of genuine instances of disease (absence of disease, respectively), which are detected as such by the diagnostic test. By contrast, the PPV (NPV) measures the proportion of cases diagnosed by the test as instances of disease (absence of disease, respectively) which are, or will turn out to be, genuine. In assessing test accuracy, it can prove misleading to focus exclusively on sensitivity and specificity.

The PPV and NPV for a disease are influenced strongly by disease prevalence (the pre-test probability that a randomly chosen person from the study cohort has the disease). The PPV increases with increasing prevalence and where prevalence is particularly low (less than 5%), the PPV can be markedly improved by moderate increases in test specificity. In interpreting a published PPV, it is essential not only to consider the CI but also to verify whether disease prevalence for the published study is representative of that for the types of patient currently under consideration. This requirement is also particularly true of the NPV.

Further, it is typically the case that an initial stage has occurred whereby diagnostic test measurements in continuous form have been classified into categories. This categorization requires the derivation of a threshold value for differentiating between diseased and non-diseased patients. The clinician may be interested in finding the threshold value that offers an optimal combination of values for sensitivity and (1-specificity). Examples of scores that have been used in this way include

- The GRACE (Global Registry of Acute Coronary Events) score in predicting death and myocardial infarction for patients with Acute Coronary Syndrome
- The APACHE (Acute Physiology and Chronic Health Evaluation) II score and GS (Glasgow Severity) score in the prediction of each of onset of severe pancreatitis, MODS (multiorgan dysfunction syndrome), and death in patients presenting with acute pancreatitis
- The MELD (Model of End-Stage Liver Disease) and UKELD (United Kingdom MELD) scores in the assessment of risk of acute liver failure and hence the prediction of waiting list mortality in patients awaiting liver transplants

The underlying procedure for deriving the threshold value involves the segregation of the test instrument scores into two groups, as determined by the gold standard, namely those who do and those who do not have the condition of interest. The accuracy of the diagnostic test is in turn assessed on the basis of these two groups. This assessment involves generating a series of threshold values and corresponding values for sensitivity and 1-specificity. The ROC curve (Receiver Operating Characteristic) involves a plot of sensitivity versus 1-specificity. If the intention is to compare the performance of competing diagnostic tests, ROC curves for the different tests can be plotted on the same graph. For any one plot, the numerically optimal combination of sensitivity and specificity values is represented by the point on the curve that is closest to the top left-hand corner. However, the trade-off between sensitivity and specificity must also be carefully weighed.

For example, if the test is confirmatory, as might be the case in human immunodeficiency virus (HIV) testing, it may be preferable to choose a slightly different point, which further reduces the proportion of false positives (1-specificity) with a small cost to sensitivity. In comparing the accuracy of two tests by means of ROCs, it is common to use the area under the curve (AUC).

Where the diagnostic test identifies cases falling into the upper (lower) range of a test score, the AUC may be interpreted as a measure of the likelihood for a randomly chosen diseased patient and disease-free patient that the diseased patient will have a higher value (lower value, respectively) than the disease-free patient.

Where ROCs do not overlap, therefore, the greater the area under the curve, the more effective the diagnostic tool. Where they do overlap, the curve with the lower overall AUC may have a peak at an optimal combination of sensitivity and specificity values not attained by the other curve. It may therefore make sense to compare the partial areas under the curves within one or more ranges of specificity values.

Further Reading

Fox, Keith A., et al. "Prediction of Risk of Death and Myocardial Infarction in the Six Months Following

Presentation With ACS: A Prospective, Multinational, Observational Study (GRACE)." *British Medical Journal* 333 (2006).

Lasko, Thomas A., et al. "The Use of Receiver Operating Characteristic Curves in Biomedical Informatics." *Journal of Biomedical Informatics* 38, no. 5 (2005).

Modifi, Reza, et al. "Identification of Severe Acute Pancreatitis Using an Artificial Neural Network." *Surgery* 141, no. 1 (2007).

Neuberger, James, et al. "Selection of Patients for Liver Transplantation and Allocation of Donated Livers in the UK." *GUT* 57 (2008).

Obuchowski, Nancy A. "Receiver Operating Curves and Their Use in Radiology." *Radiology* 229 (2003).

Pan, Jian-Xin, and Kai-Tai Fang. *Growth Curve Models and Statistical Diagnostics.* New York: Springer, 2002.

Partidge, Savannah C., et al. "Accuracy of MR Imaging for Revealing Residual Breast Cancer in Patients Who Have Undergone Neoadjuvant Chemotherapy." *American Journal of Roentgenology* 179 (2002).

Ward, Michael E. "Diagnostic Tests." www.chlamydiae.com/restricted/docs/labtests/diag_examples.asp.

Wilson, Edwin B. "Probable Inference, The Law of Succession, and Statistical Inference." *Journal of the American Statistical Association* 22, no. 158 (1927).

Zhou, Xiao-Hua, Donna McClish, and Nancy Obuchowski. *Statistical Methods in Diagnostic Medicine.* Hoboken, NJ: Wiley-Interscience, 2002.

Margaret MacDougall

Disease Survival Rates

Category: Medicine and Health.
Fields of Study: Data Analysis and Probability; Number and Operations.
Summary: Sophisticated mathematics is used to calculate disease survival rates and to help doctors and patients make treatment decisions.

Disease survival rates indicate the seriousness of a certain disease, and the prognosis of a person with the disease based on the experience of others in the same situation (in terms of the stage of the disease, gender, and age). "Overall survival rate" is defined as the percentage of people who are alive after a specific period of time after diagnosis with the disease, which is computed using the following formula: Overall Survival Rate = 100(Number alive at the end of a time period ÷ Number alive at the start of a time period).

Standard time periods such as one, five, and 10 years are often used. For instance, the five-year overall survival rate for stage-I breast cancer is said to be "95%" if 95% of all people who are diagnosed with stage-I breast cancer live for at least five years after diagnosis. Conversely, 5% of these people die within five years.

Survival rates depend on many factors, including both the type and stage of disease, as well as age, gender, health status, lifestyle, and treatment. Doctors and researchers use survival rates to evaluate the efficacy of a treatment, compare different treatments, and develop treatment plans. For example, the treatment having the highest survival rates over time is usually chosen. If treatments have similar survival rates but different numbers of side effects, the treatment with the fewest number of side effects is often selected.

Other Types of Survival Rates

Overall survival rates have some limitations. First, they do not distinguish causes of mortality within a given time period. For instance, a death may be caused by a car accident rather than by the disease. Second, they fail to indicate whether the disease is in remission or not at the end of the time period. Moreover, they do not directly provide the prognosis for a specific patient. For instance, the 95% five-year survival rate for stage-I breast cancer does not guarantee that every patient will survive more than five years. When considering only deaths caused by the disease, relative survival rate or cause-specific survival rate is often used. Relative survival rate is the ratio of the overall survival rate for people with the disease to that for a similar group of people in terms of age and gender without the disease.

One advantage is that relative survival rates do not depend on the accuracy of the reported causes of death. On the other hand, cause-specific survival rate is computed by treating deaths from causes other than the disease as withdrawals so that they do not deflate the survival rate due to the disease. When using this rate, there is no need to involve a similar group of people without the disease. Sometimes more detailed survival rates in terms of the status of a disease after a given period of time, such as disease-free survival rate and progression-free survival rate, are of interest. The computation for disease-free survival rate is similar to that of the overall survival rate except that the numerator is the number of patients who are cured at the end of the time period. Similar computation applies to the progression-free survival rate except that the numerator is the number of people who are alive and still have the disease, but the disease is not progressing at the end of the time period. As before, disease-free and progression-free survival rates can be adjusted by filtering out the effect of deaths from causes unrelated to the disease.

Survival Function

Related to survival rates, the survival function is a mathematical function that uniquely determines the probability distribution of a random variable. In survival analysis, the random variable of interest is survival time or time to a certain event, denoted by T. For instance, survival time could be time until recovery from a disease, or time to death. The survival function for T is a function of time point t defined as

$$S(t) = P(T > t)$$

which is the true probability that the survival time of a subject is beyond time t. The survival rates with an adjustment for deaths because of unrelated causes are estimates of the survival function at some t based on existing data. For a study with n patients, the survival function can be estimated by the empirical survival function:

$$S_n(t) = \text{Number of patients not experiencing the event up to } t/n.$$

In follow-up studies, however, a patient with a certain disease may withdraw, die from other causes, or still be alive at the end of the study. In such cases, the survival time T of the patient is not exactly observed but only known to be greater than a certain time (withdrawal time, death time, or time at the end of the study) called "censoring time." Then T is said to be right-censored, and the resulting set of data is called right-censored data. Based on right-censored data, the survival function can be estimated by

$$S_{KM}(t),$$

the K-M estimator developed by statisticians Edward Kaplan and Paul Meier in 1958. As a special case,

$$S_{KM}(t) \text{ coincides with } S_n(t)$$

when there is no censoring.

When estimating survival probability,

$$P(T > t) \text{ at a given time } t,$$

$S_{KM}(t)$ or a cause-of-death-adjusted survival rate introduced earlier can be used. Taking a more sophisticated approach, $P(T > t)$ can be estimated using a confidence interval. For example, one may conclude that, with 95% confidence, $P(T > t)$ is between two numbers, say 0.80 and 0.90. Such a confidence interval can be constructed using $S_{KM}(t)$ and its variance estimate from statistician Major Greenwood's formula based a normal distribution.

Further Reading

Gordis, Leon. *Epidemiology*. 4th ed. Philadelphia: Saunders Elsevier, 2009.

Kalbfleisch, John D., and Ross L. Prentice. *The Statistical Analysis of Failure Time Data*. 2nd ed. Hoboken, NJ: Wiley, 2002.

Klein, John P., and Melvin Moeschberger. *Survival Analysis: Techniques for Censored and Truncated Data*. New York: Springer-Verlag, 1997.

Marks, Harry. "A Conversation With Paul Meier." *Clinical Trials* 1 (2004).

Qiang Zhao

Doppler Radar

Category: Weather, Nature, and Environment.
Fields of Study: Algebra; Geometry; Representations.
Summary: Doppler radar uses the mathematical characteristics of waves to track and predict weather patterns.

Radio detection and ranging, commonly known by the acronym "radar," was initially developed to detect and determine the distance of enemy aircraft when visual methods were insufficient, such as in poor weather or at night. It is commonly traced to the nineteenth century work of physicist Heinrich Hertz, who investigated the reflection of radio waves from metallic objects. Doppler radar is a type of radar that uses the Doppler effect to judge the speed and direction of distant objects. The Doppler effect (also known as "Doppler shift") is a physical property that applies to all types of waves, including sound and light. Mathematician and physicist Christian Doppler presented a paper on this effect in 1842, describing how frequencies of waves change in correspondence to the relative movement between source and observer. In 1948, Hippolyte Fizeau independently discussed the shift in the wavelength of light coming from a star in similar terms. Doppler radar has applications in many fields including aviation, meteorology, sports, and traffic control. For example, Doppler radar is widely used for detecting severe weather, and it is a critical component in wind-shear detection and warning systems for airports.

Mathematics of Waves

The Doppler effect relies on the mathematical properties of waves. Transverse waves, which disturb a medium perpendicular to the direction the wave is traveling, are described in terms of their wavelength and amplitude. Wavelength is the distance between two wave crests or troughs, while amplitude is the height of the wave. An example of this is light. Longitudinal waves produce a series of compressions and rarefactions in a medium and are described by their amplitude and frequency. An example is sound, where amplitude corresponds to intensity (or "loudness") and frequency corresponds to pitch.

A car with a siren emits a series of sound waves of constant frequency. If the car moves toward a stationary observer, the waves will seem to be "bunched up" (to have greater frequency), thus a higher pitch. The same siren moving away will have waves that appear "stretched out," with lower frequency and pitch. Similarly, an oncoming light source will appear more blue, while one moving away will appear more red, corresponding to higher and lower frequencies on the electromagnetic spectrum. The amount of change in frequency is relative to both speed and direction of the moving object. The speed of a moving object can be measured by shooting waves of a known frequency at the object, and then observing the frequency of the waves that bounce from the object to the source. The difference between the outgoing and incoming frequencies is used to calculate speed. Common examples are the handheld radar guns used to measure the speed of automobiles or a thrown baseball. Edwin Hubble, for whom the Hubble Space Telescope was named, used the Doppler effect to help measure the distances to other galaxies. Light from other galaxies looks more red, indicating they are moving away. This "redshift" is commonly used as evidence in favor of the Big Bang theory of the origin of the universe.

Weather Detection

Many consider Doppler radar to be the best tool available for detecting tornadoes, hurricanes, and other extreme weather in the twenty-first century. Weather stations commonly emit radio waves that strike objects like clouds or heavy rain, and reflect back. Meteorologists use this data to determine the speed and direction of a weather system, as well as for probabilistic models to predict the path and potential severity of a storm in a given geographic area. Mathematical algorithms produce color-coded weather maps, weather animations, and other visualizations for new programs or Web sites, indicating how a storm system is predicted to move through a geographic area. Some researchers have used input data from a single radar station and knowledge of the mathematical structure of hurricanes to construct three-dimensional maps.

In the twenty-first century, a system of 21 Atlantic and Gulf coast radar stations, starting in Maine and ending in Texas, gathers real-time data to mathematically estimate the characteristics of hurricanes within 120 miles of the coast. Previously, forecasters had to fly aircraft into oncoming hurricanes and throw instruments overboard to collect data, giving them a lead time of about half a day before hurricane landfall. Other mathematicians have explored numerical weather prediction using Doppler radar and a technique known as "four-dimensional variational data assimilation," which estimates model parameters by optimizing the fit between the solution of a given model and a set of observations the model is intended to predict.

Further Reading

Harris, William. "How the Doppler Effect Works." http://www.howstuffworks.com/science-vs-myth/everyday-myths/doppler-effect.htm.

O'Connor, J. J., and E. F. Robertson. "Mactutor History of Mathematics Archive: Doppler Biography." http://www-groups.dcs.st-and.ac.uk/~history/Biographies/Doppler.html.

Schetzen, Martin. *Airborne Doppler Radar: Applications, Theory and Philosophy*. Reston, VA: American Institute of Aeronautics and Astronautics, 2006.

Sarah Boslaugh

Earthquakes

Category: Weather, Nature, and Environment.
Fields of study: Data Analysis and Probability; Measurement.
Summary: Earthquakes are measured in several ways, the most famous of which is the logarithmic Richter scale.

Earthquakes are the movements of Earth's crust resulting from tectonic plates colliding against each other. This sudden release in energy causes seismic waves that cause destruction. Depending on their severity, earthquakes range from being barely noticeable to causing permanent damage to infrastructure along with a significant loss of life. Most earthquakes are caused by the action of geological faults but they can also be caused by mine blasts, volcanic activity, and subterrestrial activity, such as injecting high-pressure water for geothermal heat capture. The focal point of the earthquake is called the "hypocenter." The point on the ground directly above the hypocenter is known as the "epicenter" of the earthquake. Philosophers, mathematicians, and scientists have long attempted to understand earthquakes. Thales of Miletus thought that earthquakes occurred because Earth rested on water. Mathematician, astronomer, and geographer Zhang Heng invented the first seismograph for measuring earthquakes in the second century. Mathematician Harold Jeffrey theorized that Earth's core is liquid after analyzing earthquake waves. Geologists use statistical methods to try to predict earthquakes.

Seismic Waves
A tremendous amount of energy is released from the epicenter radially outward. As the energy spreads, it is manifested in three forms: compression waves (P waves), shear waves (S waves), and surface waves.

P waves are felt first and do minimal damage. S waves follow the P waves and do minimal damage. It is the slower surface waves (also known as "Love waves") that cause the majority of the damage.

Measurement
The goal of earthquake measurement has been to quantify the energy released. Seismographs are highly sensitive instruments employed to record earthquakes. Conventionally, earthquake magnitudes are reported in the Richter scale. The Modified Mercalli Intensity Scale is commonly used to ordinally quantify (or rank) the effects of an earthquake on humans and infrastructure. Body wave or surface wave magnitudes are also used to measure earthquakes.

Richter Scale
The Richter scale quantifies the amount of seismic energy released during a quake. It is a base-10 logarithmic scale, which means that the difference between an earthquake of rating 2.0 on the Richter scale and 3.0 correlates to a tenfold increase in measured amplitude. Specifically, the Richter scale is defined as

$$M_L = \log_{10} A + B$$

where A is the peak value of the displacement of the Wood–Anderson seismograph (mm) and B is the correction factor. The wave intensity measurements are also logarithmic functions, using variables such as the ground displacement in microns, the wave's period in seconds, and distance from the earthquake's epicenter.

Modified Mercalli Intensity Scale
The Modified Mercalli Intensity Scale has 12 gradations: instrumental, feeble, slight, moderate, rather strong, strong, very strong, destructive, ruinous, disastrous, very disastrous, and catastrophic.

Further Reading
Brune, James. *Tectonic Stress and the Spectra of Seismic Shear Waves From Earthquakes*. La Jolla, CA: Institute of Geophysics and Planetary Physics, UCSD, 1969.

Gutenberg, B., and C. F. Richter. *Earthquake Magnitude, Intensity, Energy and Acceleration*. Pasadena, CA:

Bulleting of the Seismological Society of America, 1956.

Hough, Susan. *Predicting the Unpredictable: The Tumultuous Science of Earthquake Prediction.* Princeton, NJ: Princeton University Press, 2009.

Ashwin Mudigonda

Elementary Particles

Category: Space, Time, and Distance.
Fields of Study: Data Analysis and Probability; Number and Operations; Representations.
Summary: Various branches of mathematics are employed to study elementary particles, the smallest particles in the universe.

Particle physics is a branch of physics that seeks to describe and explain the universe on the smallest scales. The particles thought to be the fundamental building blocks of matter and force are called "elementary particles." Like all branches of physics, the study of elementary particles relies heavily upon many branches of mathematics, including calculus, geometry, group theory, algebra, and statistics. Particle physics also contributes to mathematical research by posing questions that give rise to new mathematical theories.

History

For thousands of years, scientists and philosophers have been asking the questions, "What is the universe made of?" and "Are there fundamental units that make up space, matter, energy, and time, or are these infinitely divisible?" As early as the fifth century b.c.e., Greek philosopher Democritus (c. 460–370 b.c.e.) hypothesized that all matter is made of indivisible, fundamental units called "atoms." Despite these early hypotheses, there was very little progress in this field until the dawn of the twentieth century.

The twentieth century saw the emergence of several new branches of physics. Among these was particle physics, a field that seeks to explore the universe on the smallest scales. Particle physicists try to identify the particles that form matter and force, describe their properties, and understand how these particles relate to each other. Some of these particles are not composed of any other particles and are therefore called "elementary particles." These elementary particles form the basic building blocks of the universe.

The understanding of particle physics at the beginning of the twenty-first century is embodied in the Standard Model of Particle Physics, an elaborate yet still incomplete model that attempts to list and describe all existing particles. Jokingly referred to as "The Particle Zoo," the Standard Model lists dozens of particles and includes elementary particles with exotic names such as "gluon," "muon," and "quark." Many of the particles in the Standard Model have yet to be detected experimentally, and their existence is conjectured based on theoretical work.

Mathematics Used in the Study of Particle Physics

Like all physical theories, particle physics relies heavily upon mathematics, which provides the theoretical framework physicists use to explain and describe physical phenomena. Mathematics also enables physicists to make predictions that can later be tested using modern tools, such as particle accelerators.

One of the most useful branches of mathematics is calculus, a field that has applications in practically all branches of the natural sciences, as well as in engineering and even in the social sciences. It is therefore not surprising that calculus occupies a central role in the theory of elementary particles. Differential calculus may be used to describe properties of particles at an instant, while integral calculus is used to describe cumulative effects of a particle or a system of particles over time and space.

Calculus is but one branch of the mathematical field of analysis that is useful in particle physics. Other branches of analysis—partial differential equations, complex analysis, and functional analysis—play important roles as well.

Geometry has traditionally been used to describe the universe on the grandest scales, those of galaxies, galaxy clusters, and the universe as a whole. Recently, geometry has found a place in elementary particle research as well. French mathematician Alain Connes (1947–) has described a theoretical model for particle physics that is based on noncommutative geometry, which is a geometrical representation of noncommutative algebras—systems in which the order of factors in an operation determines the value of the operation. For example, if a and b are real numbers, then it is always true that $a \times b = b \times a$, as multiplication is commutative for real numbers. However, if A and B are matrices, then generally $A \times B \neq B \times A$. Matrix multiplication is therefore noncommutative.

Symmetry, Group Theory, and Quantum Mechanics

One of the most fundamental mathematical concepts in elementary particles is symmetry. In mathematics, symmetry is defined as an operation on an object that leaves some of the object's properties unchanged. As an example, consider a square drawn in the plane and an axis of rotation that passes through the square's center, perpendicular to the plane. If the square is rotated by 90 degrees around that axis, the square will appear unchanged. Rotation by 90 degrees is thus called a "symmetry" of the square. The set of all symmetries of an object forms a mathematical construct called a group (a set with an operation that obeys several axioms). Group theory, a branch of algebra, plays an important role in particle physics, as properties of many elementary particles can be explained and described by the use of symmetry.

The chief group-theoretic structure in particle physics is the Lie (pronounced "Lee") group, named after Norwegian mathematician Sophus Lie (1842–1899). Lie groups are groups that posses the properties of

NASA scientists detected a ring of dark matter that formed during a collision between galaxy clusters. Astronomers don't know what dark matter is made of; however, they believe it is a type of elementary particle. (National Aeronautics and Space Administration)

geometric constructs known as "differentiable manifolds." Lie groups thus provide yet another connection between geometry and elementary particles.

One of the most important physical theories of the twentieth century is quantum mechanics, a theory that holds that, at the atomic and subatomic levels, behavior of particles is a statistical rather than a deterministic phenomenon. Since elementary particles obey quantum-mechanical laws, statistics and probability are invariably major components of the mathematical framework of elementary particles.

While physicists use mathematics as a tool for exploring the universe, the relationship between particle physics and mathematics is not one-directional. Research in particle physics drives the emergence of new mathematical theories, just as mechanics drove the emergence of calculus in the seventeenth century. In 1990, American theoretical physicist Edward Witten (1951–) won the Fields Medal, the highest honor in mathematics, for his many contributions to mathematics. He is the only non-mathematician ever to win the prestigious award. As both mathematicians and physicists continue to explore new horizons, the cross-fertilization of ideas will benefit both fields in decades to come.

Further Reading
Griffiths, David. *Introduction to Elementary Particles*. Weinheim, Germany: Wiley-VCH, 2008.
Hellemans, Alexander. "The Geometer of Particle Physics." *Scientific American* 295, no. 2 (2006).
Mann, Robert. *An Introduction to Particle Physics and the Standard Model*. Boca Raton, FL: CRC Press, 2010.

Or Syd Amit

Elevation

Category: Space, Time, and Distance.
Fields of Study: Geometry; Number and Operations.
Summary: Various aspects of elevation can be calculated using mathematical techniques.

Trigonometry has long been used to measure height. Elevation is often the height of a point relative to sea level, and its measurement is called "hypsometry." Elevation affects air pressure, temperature, and gravity, all of which have noteworthy effects on people. Astronomers and mathematicians such as Blaise Pascal and Edmund Halley investigated relationships between barometric pressure and elevation.

Historical surveys of elevation include those who used barometers, like John Charles Frémont, who was at one time professor of mathematics of the Navy, and physician Christopher Packe. However, this method is sensitive to a number of variables. In the twenty-first century, detailed elevation data are available. Mount Everest is known as Earth's highest elevation. Topographical maps represent elevation by using contour lines, each line following a path of constant elevation. Transits were developed in the nineteenth century, and they can be used to calculate changes in elevation. Contour integrals and generalized contours for functions of two variables are investigated in multivariable calculus classrooms. Mathematicians and computer scientists have helped create realistic computer models of land elevation, called "digital elevation models." They have explored ideas like irregular-mesh grids or shifting nested grids in surface reconstruction. Other types of elevation studies also benefit from mathematical techniques, like using the ocean wave spectrum to investigate sea surface elevation peaks, or statistical techniques to investigate the impacts of elevation changes. Mathematician and astronomer Nilakantha Somayagi investigated the elevation of lunar cusps in the sixteenth century. The term "angle of elevation" in high school classrooms represents the angle between where an observer is standing and the line of sight to an object. The angle of elevation is found in many contexts, including in the Pyramids of Egypt, in the astrolabe, and in global positioning systems.

Topographic Maps
A topographic map is a two-dimensional map that conveys elevation information as well as other features of an area. Contour lines are the key to capturing elevation changes from a three-dimensional world on a two-dimensional map. A contour line is a path that follows a constant elevation. Early uses of contours date to the eighteenth and nineteenth centuries and include the work of engineer Jean-Louis Dupain-Triel and astronomer and mathematician John Couch Adams.

A contour line is drawn each time a predetermined elevation change is achieved. For example, a map may use 100-foot elevation increments, with one contour line following points having an elevation of 100 feet and the next marking an elevation of 200 feet. Consecutive contour lines always differ by 100 feet in elevation. As the mapped terrain climbs more steeply, the contour lines on the map will be closer together. The lines can mark elevations that increase and decrease, representing terrain that rises and falls intermittently. Contour lines can represent elevations that are zero, or negative numbers as when mapping an ocean floor.

A topographic map of an area with constant elevation at its boundary, such as an island bounded by the sea, will not have contour lines extending off the map's edge. In such cases, all contour lines will appear as closed curves. A curve is closed if it loops back to where it started. Typically, contour lines appear as simple closed curves that do not cross themselves. The pattern of contour lines as nonintersecting rings lying one within another is common on topographic maps. Also common is to have two separate sets of nonintersecting rings contained within a single contour line, as when two hills are surrounded by a larger path of constant elevation.

The U.S. Geological Survey (USGS) has created a complete large-scale topographic map of the United States in more than 56,000 pieces. The National Elevation Dataset is noted as the "the primary elevation product of the USGS." The data set is updated regularly, and historic data sets are also available for investigations.

There is an ever-growing growing need for digitized maps, which allow a computer user to read elevation at any spot on the map. Some digitized maps enable the user to view a landscape from different perspectives, creating a three-dimensional view of the area's elevation changes, similar to what would be seen at the actual location. Data from existing topographic maps and aerial photography are used to create digitized maps. Improvements in technology will continue to affect the science of map making.

Effects of High Elevation

As elevation increases, air temperature drops because of a decrease in air pressure. At about 18,000 feet above sea level, for example, the air pressure is half that at sea level. In the troposphere, the lowest layer of Earth's atmosphere, a general rule of thumb is that air temperature drops 6.5 degrees Celsius for every 1000 meters of elevation gain, or roughly one degree Fahrenheit for every 280 feet of elevation gain in standard conditions. This phenomenon, which can be modeled with an equation, can be seen directly when an observer standing at a low elevation on a warm day views a tall mountain covered with snow.

(iStockphoto)

Highest Elevations on Earth

Elevations are nearly always computed relative to sea level, the average height of the ocean's surface. Sea level is an inexact measure since tides, temperature, wind, salinity, and air pressure affect the oceans. Mount Everest (above) in the Himalaya Mountains near the border of Nepal and Tibet is the highest mountain on Earth at an elevation of 29,035 feet as of 2010. Everest gains more than two inches of elevation per year because of the collision of tectonic plates and there are discrepancies in its listed height.

Earth is not spherically symmetric; its radius near the equator is more than 13 miles greater than its radius near the poles. Consequently, Mount Chimborazo in Ecuador holds the distinction of having the summit farthest from Earth's center. Lying about one degree south of the equator, where Earth is widest, Mount Chimborazo is approximately 20,561 feet above sea level, enough to make its summit more than a mile farther from Earth's center than Mount Everest's summit.

Another consequence of this cooling is that water vapor in the air condenses, sometimes causing increased rainfall on the windward side of a mountain range and a "rain shadow" downwind from the mountains. Many deserts lie just downwind from a mountain range. For example, sand dunes in Death Valley, California, lie in the rain shadow of Mount Whitney, the highest peak in the continental United States.

Because of these differences in temperature and precipitation, tall mountains can have multiple climatic zones, with different plant species thriving near the summit than at lower elevations. Some animal species, such as Roosevelt elk, migrate seasonally to take advantage of elevation effects, climbing to cooler locations in the summer and descending to warmer valleys in winter.

The lower atmospheric pressure at high elevations makes breathing more difficult. Mountain climbers at high elevations use special apparatus to breathe. Some competitive distance runners train at high elevations in order to challenge their cardiovascular systems. When they race at a lower elevation, the air feels relatively dense and oxygen-rich, giving them a competitive advantage.

With the less-dense atmosphere at high elevations, the sun's rays can penetrate more easily, making sunburn possible even on cold days. Engines of naturally aspirated cars get less horsepower at higher elevations. Projectiles travel farther, a phenomenon known to golfers and baseball players. Standard equations for projectile motion sometimes assume a sea-level location; adjustments must be made to account for elevation.

The effect of gravity is reduced with travel to high elevations; mass remains the same but weight decreases slightly, primarily because of the increase in distance from Earth's center of mass. A person's weight would be less atop Mount Chimborazo than anywhere else on Earth.

Further Reading

Smith, Arthur. "Angles of Elevation of the Pyramids of Egypt." *Mathematics Teacher* 75, no. 2 (1982) .

Thrower, Norman. *Maps & Civilization: Cartography in Culture and Society*. 3rd ed. Chicago: University of Chicago Press, 2008.

U.S. Geological Survey. "National Elevation Dataset." http://ned.usgs.gov.

David I. Kennedy

Energy

Category: Space, Time, and Distance.
Fields of Study: Algebra; Measurement.
Summary: Mathematics is used to study energy and energy conservation as well as to develop new sources of energy.

The concept of energy and transportation of energy are central to the survival of any civilization. As mathematical physicist Ludwig Boltzmann noted, "Available energy is the main object at stake in the struggle for existence and the evolution of the world." At the start of the twenty-first century, human beings have accessed or created many forms of energy and power production, including coal-fired and oil-fired power plants, solar heating plants, wind farms, nuclear power plants, geothermal sources of heat, hydroelectric power produced by dams, biofuels that store solar energy, and tidal energy produced by gravitational interactions between Earth and the moon.

There are also potentially disruptive energy sources, including natural events, such as lightning, volcanoes, and earthquakes. Some global sources of energy and power that remain to be tapped by humans include the atmosphere's expansion and contraction, ocean currents, and sea level differences. Various calculations of energy, including chemical reactions and nuclear reactions, invoke the principle of conservation of energy. In relativistic or quantum terms, the conservation of mass-energy is also important. Energy, work, and quantity of heat are all expressed in "joules," a measure of work named for physicist James Joule. There is a vast array of energy problems that mathematicians research, and mathematics makes many contributions to energy issues.

Energy, Defined

Energy is found in nearly every system or process in the universe: mechanics, chemicals, heat, electricity, nuclear processes, and quantum effects. Mathematician and scientist René Descartes studied mechanics; centuries later, mathematician and philosopher Gottfried Leibniz criticized his ideas and developed what are referred to today as "kinetic energy," "potential energy," and "momentum." In mechanics, the kinetic energy (E) of an object is expressed as

$$E = \frac{1}{2}mv^2$$

where *m* is the object's mass and *v* is its velocity. Another form of energy found in mechanics is the energy of position called "potential energy." It has the units of joules. An example is the potential energy defined as work done in the compression of a coiled spring. The sum of all the kinetic and potential energies within a system comprises the mechanical energy of the system. Energy may be a conserved quantity within a closed system, or it may change forms, such as mechanical energy being converted to heat by friction. How energy in a system is measured is important. As noted, mechanical energy is measured as the sum of kinetic energy and potential energy, or energies of motion and position. Chemical energy is measured by the heat energy released in chemical reactions. Electrical energy is measured by work done in a system.

Energy Conservation

In general, the amount of energy of various types can be equated to an equivalent amount of heat energy. On an experimental scale, heat energy is the ability of work done to raise the temperature of water. The joule is a measure of thermodynamic energy and is the common unit of energy. James Joule is credited with experiments in the mid-1800s that demonstrated that work done on a system can be converted into heat. His experiments and those of others eventually led to the realization and statement of the "principle of conservation of energy" as a hypothesis, which was proved in certain restricted settings and generalized by induction. In 1865, mathematical physicist Rudolf Clausius worked on thermodynamics and stated his first law as, "The energy of the universe is constant." The principle of conservation of energy applies not only to certain mechanical systems but is also seen widely in systems where other forms of energy are considered. Thus, heat energy is produced by combustion and friction, radiant energy is from light and other forms of radiation, and chemical energy is stored in fuels and electrical energy. The principle is continually tested in new situations. This testing led to discoveries in the twentieth century in atomic physics. In the International System of Units, *Le Système International d'Unités* (SI), a joule is defined as a newton-meter, named for Isaac Newton. The systematic study of the relation of various physical quantities through an analysis of their dimensions is the subject of dimensional analysis. Richard Feynman noted, "For those who want some proof that physicists are human, the proof is in the idiocy of all the different units which they use for measuring energy."

One energy issue that has been important to mathematicians, philosophers, and physicists is the relationship between matter and energy. Some physicists wanted to assign matter-like properties to energy, such as Wilhelm Wien, who considered that energy might have a traceable motion. Mathematician William Clifford thought of matter and energy as types of curvatures. In the theory of special relativity of 1905, Albert Einstein proved an equivalence of mass and energy as expressed in his famous equation $E = mc^2$, where E is the energy equivalent of mass m, and c denotes the speed of light, 299,792,458 meters per second. There is no process available to human beings at the start of the twenty-first century in which matter can be converted completely into radiant energy.

For example, in a nuclear explosion, only a tiny fraction of nuclear material is converted into energy. The only known process of annihilating matter is to pair a particle of matter with a particle of anti-matter, with the result that two photons are formed with energies that are equivalent to the energies of the particles. This process is on a quantum scale. Fusion is one process for partially converting mass into energy and occurs naturally in stars. Many controlled fusion experiments have been performed but in the process of producing fusion, a greater amount of input energy is needed for the reaction than is ultimately released by the reaction. Only in uncontrolled thermonuclear explosions are large amounts of energy released by fusion.

Fusion

Scientists continue to explore novel sources of energy and power from sources that entail motion, heat, quantum uncertainty and other natural physical phenomena. One possible source of power is controlled fusion reactions, hot or cold. Controlled hot fusion reactions have not yet reached a break-even point where the energy of the reaction exceeds the energy input needed to trigger the reaction.

There are ongoing fusion experiments that use various solids and liquids with energy pumped into them by lasers in which fusion occurs but the fusion is not self-sustaining. The main problem is the energy input and inherent danger in heating suitable substances to temperatures at which fusion between atoms of hydrogen isotopes can occur. The hydrogen is in the form of deuterium or tritium, and the temperatures reached through compression must be on the order of millions of degrees, and there are often energetic byproducts that are dangerous to objects and people. In contrast to hot fusion, cold fusion (also known as "low-energy nuclear reactions" among the twenty-first-century research community) is the fusion of atoms at close to room temperature, generally through the use of supersaturated metal hydrides. These reactions produce heat, helium, and a very low level of neutrons. The energy output is greater than the input, leading many scientists and others to investigate this process as a viable solution to the energy needs of the future. Chemists Martin Fleischmann and Stanley Pons were the first, in 1989, to publicly announce that they had achieved cold fusion. Many competing scientific and mathematical models have been developed to explain how cold fusion works but many researchers and others remain skeptical regarding its existence or viability.

Other Mathematical Applications

Mathematicians and other scientists have long studied the various aspects of energy. The concept of energy is fundamental to many scientific and business theories, applications, and disciplines. For instance, mathematicians have modeled energy trading in financial markets, which is quantitatively interesting because, in such applications, energy possesses unique attributes as a non-storable and non-fungible commodity. They have also worked to design efficient shutdown schedules for electronic systems to address concerns related to energy conservation. Mathematics is important for explaining the cosmic phenomenon of dark energy. This type of energy, often modeled as a scalar field and inferred in large part from observation and mathematical analysis of gravitational fields, has implications for theories and measurement of universe expansion and dark matter. On the other hand, mathematicians such as Blake Temple have used mathematics to attempt to disprove the existence of dark energy and posit alternative explanations. Others have investigated the geometry of symplectic energy. Mathematicians are also influential in energy research and policy making via work at federal agencies like the U.S. Department of Energy. Mathematician J. Ernest Wilkins was a fellow at the Department of Energy's Argonne National Laboratory and physicist and mathematician Hermann Bondi was the chief scientific adviser to the Department of Energy. Mathematical analysis and computational methods have also been used to study energy problems related to equilibrium, stability, and energy transport.

Further Reading

Coopersmith, Jennifer. *Energy, the Subtle Concept: The Discovery of Feynman's Blocks from Leibniz to Einstein*. New York: Oxford University Press, 2010.

Gerritsen, Margot. "Mathematics Awareness Month—April 2009—Theme Essay: Mathematics in Energy Production." http://www.mathaware.org/mam/09/essays/Margot_EnergyMaths.pdf.

Greengard, Claude, and Andrzej Ruszczynski. *Decision Making Under Uncertainty: Energy and Power*. New York: Springer, 2010.

Society for Industrial and Applied Mathematics. "Fuel Cells, Energy Conversion, and Mathematics." http://www.siam.org/about/news-siam.php?id=1605.

Veigele, William. *How to Save Energy and Money at Home and on the Highway: The Mathematics and Physics of Energy Conservation and Reduction of Consumer Energy Costs*. Boca Raton, FL: Universal Publishers, 2009.

Julian Palmore

Extinction

Category: Weather, Nature, and Environment.
Fields of Study: Algebra; Problem Solving.
Summary: Causes and factors of extinction can be quantified and modeled using mathematical and statistical techniques.

Extinction occurs when the last member of a species dies. A species survives for much longer than any of its members. For example, a human can live up to about 120 years, whereas the human species (*Homo sapiens*) is thought to have existed for hundreds of thousands of years. It is not known how long our species will endure and indeed most species on Earth have already become extinct. There are many causes of extinction, some natural and others as a result of human activities. Many factors influence whether an endangered species can avoid extinction. These factors can be quantified and modeled using mathematical and statistical techniques. A species can disappear in some parts of its habitat but not in others. Not all species have existed on Earth for the same length of time—some appear only briefly while others manage to persist for incredibly long periods of time. Human activities may be increasing the rate at which other species become extinct.

Rise of Extinction

A species is endangered when it consists of a small number of members. In such cases, individuals may have trouble finding each other because of geographical separation. For a species that is endangered, it is of interest to know whether the species is likely to become extinct. It is customary to let $N(t)$ represent the size of a population at time t. The fact that the species is endangered implies that $N(t)$ takes positive values close to zero. If $N(t)$ is eventually measured to be zero, then the species has become extinct. However, if $N(t)$ rebounds to larger positive values, then the species persists. In general, stochastic effects largely determine whether an endangered species will become extinct. Given population data $N(t)$ at different times t, one may compute the mean (μ) of the population growth rate.

$$R(t) = \ln\left(\frac{N(t)}{N(t-1)}\right).$$

For example, if $t = 10$ then

$$\mu = \frac{\left(R(1) + R(2) + \cdots + R(10)\right)}{10}.$$

A positive (or negative) value of μ indicates that the population is growing (or declining) on average. Combining this information with the standard deviation (σ) of $R(t)$ allows one to assess the risk for extinction, which is typically highest when μ is negative and σ is small. Complex models of population dynamics exist to predict whether a species will persist or become extinct. These include geometric growth models in which a population multiplies at a fixed rate, logistic growth models in which populations slowly attain steady-state sizes, and Lotka–Volterra predator-prey models for interactions between multiple species, named for Alfred Lotka and Vito Volterra.

Local Extinction

A species can become extinct in one area (such as an island) and still persist elsewhere (such as a continent). If the species is able to recolonize the former area, then this is known as a "rescue effect." If local extinction events become synchronized—as a result of global climate change, for example—then the risk of a species becoming globally extinct is much higher.

Rate of Extinction

Scientists estimate that there may be 10 million species alive today and yet they account for fewer than 1 in 1000 species that have ever lived. The average time to extinction for a species, as measured from the time of its first appearance, is close to 10 million years. When the time to extinction for a species is much longer, such as more than 100 million years, then later members are said to be living fossils.

Causes of Extinction

A species can become extinct for various reasons, including intense competition with other species, disease, or failure to adapt to changing climatic conditions, as well as the disappearance of a species' prey. Anthropomorphic reasons for extinction include overhunting by humans, habitat loss from human activities such as deforestation, and social planning (the intentional eradication of smallpox).

(iStockphoto)

Mass Extinction

A mass extinction occurs when a large number of species become extinct in a short period of time. Although rare, the fossil record indicates that these events have occurred at least five times, the most famous being the mass extinction of non-flying dinosaurs 65 million years ago in what was probably a meteor impact. Many scientists believe that we are currently in the midst of a sixth mass extinction, with up to 40,000 species becoming extinct each year—a rate that is roughly 100–1000 times higher than in prehistoric times.

Further Reading

Allen, Linda J. S. *An Introduction to Mathematical Biology*. Upper Saddle River, NJ: Prentice Hall, 2007.

Bright, Michael. *Extinctions of Living Things* (*Timeline: Life on Earth*). Portsmouth, NH: Heinemann, 2008.

Erickson, J., and A. E. Gates. *Lost Creatures of the Earth*. New York: Facts on File, 2001.

Hallam, T. *Catastrophes and Lesser Calamities: The Causes of Mass Extinctions*. New York: Oxford University Press, 2005.

Hecht, J. *Vanishing Life: The Mystery of Mass Extinctions*. New York: Atheneum, 2009.

Thieme, Horst R. *Mathematics in Population Biology*. Princeton, NJ: Princeton University Press, 2003.

Andrew Nevai

Farming

Category: Weather, Nature, and Environment.
Fields of Study: Algebra; Geometry; Measurement.
Summary: As a fundamentally important human activity, agriculture has long been a motivator for mathematical and statistics research.

Farming, also called "agriculture," is the production and distribution of plant and animal products. Farming methods range from organic farming to industrial agriculture. Farming operations are also categorized by their products, including foods, pets, decorative plants, pharmaceuticals, building materials, fibers, resins, and bioplastics. Agriculture has long been a motivator for mathematical and statistic research. Mathematical concepts and models have helped advance many agricultural methods beyond simple arithmetic calculations of quantities of seed and fertilizer. Many consider Ronald Fisher to be the father of modern statistics. Much of his research in statistical methods originated from his work with more than 60 years of agricultural data at Rothamsted Experimental Station, one of the oldest agricultural research institutions in the world. Methods pioneered by Fisher are still widely used in the twenty-first century, including hypothesis testing, analysis of variance, maximum likelihood estimation, and factorial experimental design. Mathematician Michael Weiss has worked in several mathematical areas with applications in agriculture, including nonlinear and chaotic dynamics, fuzzy set theory, and topological and algebraic entropy. Some applications of his work include a model of crop yields as a two-dimensional stochastic process, called "random surfaces," and assessing revenue risk as a probabilistic function of foodborne disease outbreaks. Precision farming models spatial variability in farmland and the resulting changes in yields as geometric surfaces.

Numerical characteristics of the farmland, such as fertilizer needs, are assigned to surfaces by functions and mapped to other surfaces by operators using modeling software. The so-called cobweb theorem relates price and production for situations in which there is a time lag between the marketing of a product and initially obtaining price information to determine production. This is common in agricultural markets, since prices in one year tend to influence planting in subsequent years.

The Role of Agriculture in the History of Mathematics and Science

Agricultural development shaped the history of humankind, including the growth of science in mathematics. This impact is acknowledged in the historical tradition of naming major farming breakthroughs "revolutions," since the changes they produced in society were large and relatively fast. The neolithic revolution started circa 8000 B.C.E. and included the development of permanent settlements. The resulting architecture and centralized management systems required abstract thought and systems of knowledge, including writing, mathematics, and science. The Arab agricultural revolution took place in the eighth through the thirteenth centuries C.E. and included the development and distribution of international knowledge exchange, sophisticated algebra and geometry, and astronomy for farming and navigation, as well as the scientific method and the modern number and computational system in mathematics. The British agricultural revolution started in the seventeenth century. It codeveloped with the Industrial Revolution and included the heavy use of mechanical tools and developments in the natural sciences, including chemistry and biology. This industrialization of agriculture continued into the twentieth century, driving the research in organic chemistry and genetics known as the green revolution.

Domestication of local crops, such as rice in China around 8000 B.C.E., allowed for both population growth and population concentration in villages and, later, towns. Planting, harvesting, and other timed activities required relatively exact time and weather observation, which in turn led to the development of astronomy and the development of sophisticated time measurement tools and calendars. Circa 5000 B.C.E., the people of Mesopotamia employed intensive farming methods, including monocrop fields, aggregation of crops for trade, and complex irrigation. Such methods called for and enabled major technological developments, such as better plows. It is hypothesized that the complex division of labor, distribution, and observation of water levels and calendars required for this type of agriculture led to the development and relatively widespread use of writing.

Mesopotamian clay tables show that quadratic and cubic equations, the Pythagorean theorem, and other topics currently found in algebra, geometry, and calculus were already widely used circa 2000 B.C.E. in problems related to agriculture, such as astronomy-based calendars to time flooding and harvesting or the distribution of products. Some of this knowledge later was lost and then rediscovered by other cultures, and some continued to be used in the original form. For example, the practice of measuring time based on 60 minutes in an hour and 60 seconds in a minute comes from the Babylonian sexadecimal (base 60) number system. The number "60" was a convenient one for the Babylonians being highly composite (with more divisors than any number less than 60).

Agriculture promoted the development and spread of increasingly complex mechanisms, such as waterwheels in China. Excess crops supported the development of trade and transportation, from the domestication of draft and pack animals in ancient times to sophisticated spice trade fleets circa the sixteenth century. Starting in the eighth century C.E., Muslim traders established an extensive network of trade routes among Asia, Europe, and Africa, enabling the diffusion of agricultural techniques and crops beyond their places of origin. This Arab agricultural revolution led to the development and distribution of science and mathematics, including the Arabic numerals used around the world in the twenty-first century. For example, one of the first documented uses of the scientific method comes from thirteenth-century work on medicinal plants and agronomy (the farming of plants).

The Industrial Revolution, starting in the eighteenth century, included the increasing mechanization of agriculture. Agricultural machines, such as the tractor, both decreased the number of people required for farming and increased productivity. The scientific advances associated with these developments primarily took place in engineering and chemistry. The green agricultural revolution of the second half of the twentieth century promoted advances in chemistry, genetics, and bioengineering, which led to high-yield, disease- and pest-resistant cultivation of major crops. The sustainability of these practices is not yet clear at the start of the twenty-first century.

Measurements in Agriculture

Metrics used in farming focus on average production of different cultivars of plants, breeds of animals, or farming methods; resource intensity of practices; efficiency of distribution; nutritional value of food products and industry-specific values of fibers, fuels, and lumber; environmental impact and sustainability; and the role of agriculture in local and global economy.

The global production levels, by crop type, are measured in tons per year. For example, cereals was the number one category of agricultural product, with worldwide production at around 2 billion tons per year in the early twenty-first century, while meat production at this same time was around 250 million tons per year. The total and per capita rates of production are frequently compared between years. For example, the total agricultural production grew by a factor of 16 between the early 1800s and 1970, while the world population grew by a factor of seven. This means that per capita consumption of agricultural products more than doubled during that period but not necessarily because of food items. Fiber or farmed trees for paper and construction are also included.

Farm yields are measured in crop weight per area for plants; in the ratio of seed input to seed output for grains; or in meat, fiber, or egg production per animal for animals. The yields are estimated using statistical methods of random sampling, or total outputs of a farm. In the United States, for example, corn yields averaged about 30 bushels per acre in the early 1900s and around 130 bushels per acre in the early 2000s. Food anthropologists estimate the minimal ratio of grain input to output necessary for sustaining farming as the main source of food as 1:3. For each grain planted, farmers get three grains, one of which is planted and two of which are either eaten by people or fed to farm animals. Yield metrics can be used to compare different methods of farming. For example, irrigation can raise corn yields by a factor of four or five. Industrial farming in developed countries produces yields that are about 10% greater than organic farming in nondrought years and about 70% less in drought years, netting about the same average yields over decades.

Resource intensity is measured by the outside input required per area of crops, per individual animal in meat or egg farms, or per unit of farm product output. For example, it takes about 1000 liters of water to produce 1 liter of corn-based ethanol. Resource intensity is one of many sustainability metrics used in farming. Other mathematical metrics of sustainability include nutrient leaching into water systems, which may cause proliferation of algae; biodiversity of farms; and pollution of soil, water, and air with herbicide and pesticide residues; as well as the carbon footprint of farming practices. For example, livestock production is currently responsible for about one-fifth of the total carbon footprint of humanity.

Farming and the Economy

Agricultural systems include production, processing, packaging, distribution, marketing, and consumption. The proportion of resources and energy required for these activities varies with farming practices. For example, eating local foods reduces the resources expended in transportation; operating monocrop farms reduces labor per unit of production; eating processed foods increases packaging costs.

Agricultural economics is the study of resource allocation and distribution related to agriculture. It uses mathematical statistics for data analysis and trend prediction and mathematical modeling for research and development. Many general economic mathematical models were first developed in agricultural economics, for example, the cobweb model, which explains the cycles of price fluctuations through analyzing lags within the production chains, such as planting and harvesting.

Combines harvesting crops at precise intervals with each row overlapping slightly. Combines were invented in 1834. (iStockphoto)

Factory farming uses economies of scale by raising livestock in confinement and with high population densities. The calculations involved in factory farming include cost-output analysis and bioengineering of animals to optimize product output as well as the logistics of supplying food in to each animal in place and disease prevention through administering antibiotics. There are several measurements of factory farming impacts. For example, there are metrics involved with animal welfare, such as the degree of confinement, measured in area of pen per animal. Human health impact measures and research include studies of pesticide, antibiotic, and growth hormone levels in farm products and statistical studies of the impact of food on human health. Environmental impact measures are standard for all operations and include levels of specific air, water, and soil pollutants produced by the farm and its carbon footprint. Capital redistribution is the measure of movement of money among communities, which is relatively high for factory farming because of its centralized nature.

Industrial marketing and distribution models do not work well for organic farming because most organic products are not scalable. In the early 2000s, organic farmers developed a variety of peer-to-peer credence and distribution models, network marketing models, and sharing economy (mesh) models. Such modern models support decentralized production and disintermediated distribution. Some organic farmers join together in cooperatives and use economies of scale. Community-supported agriculture (CSA) is an economic model that provides a way to share the benefits and risks of farming. In a typical CSA, consumers buy farm shares and receive a weekly delivery of farm outputs.

Further Reading

Alspaugh, Shawn. "Farmer Ted Goes 3D." *Mathematics Magazine* 78, no. 3 (2005).

Glen, John. "Mathematical Models in Farm Planning: A Survey." *Operations Research* 35, no. 5 (1987).

Street, Deborah. "Fisher's Contributions to Agricultural Statistics." *Biometrics* 46 (1990).

Weiss, Michael. "Precision Farming and Spatial Economic Analysis: Research Challenges and Opportunities." *American Journal of Agricultural Economics* 78, no. 5 (1996).

Maria Droujkova

Fertility

Category: Medicine and Health.
Fields of Study: Algebra; Data Analysis and Probability.
Summary: Individual fertility cycles can be mathematically predicted and national fertility rates are a useful statistical measure for analyzing population demographics.

The term "fertility" has been used historically in a variety of contexts, including the richness of croplands with respect to producing food, the creativity of the human mind and imagination, and the ability of people to have children. The term "fecundity" is often interchanged with fertility when discussing human reproduction. However, nineteenth-century physician Matthews Duncan, who researched birth statistics and fertility, differentiated the two terms by defining fecundity in an essentially binary fashion as the capability of bearing children or not, versus fertility, which he used to quantify the number of children a woman had borne. Demographers often use fertility rate as a standardized metric to describe the number of children borne per person, couple, or population and to make comparisons across populations. Many collections of global statistics, like the *CIA World Factbook*, include fertility rates, which have been connected by mathematical and statistical models to economic measures such as individual income or a country's gross domestic product. Others study relationships to medical and social variables, such as the availability of birth control and assisted reproduction or attitudes about single parenting. Some rates adjust for women in specific age groups or other variables. At the start of the twenty-first century, organizations such as the United Nations also began to turn serious attention to the issue of population decline in many nations and its potential effects on national economies, workforces, and social security systems. Mathematicians, statisticians, demographers, and others continue to research the reciprocal relationships between fertility and other measures to attempt to determine causes and effects and to forecast future trends as well as to contribute to the development of technologies related to fertility and reproduction. Statistician Leslie Kish was awarded the American Statistical Association's Samuel S. Wilks Award for his work on the World Fertility Survey, which "illustrates his impact as an international ambassador of statistics and a tireless advocate for scientific statistical methods."

Fertility Rates

In the years immediately following World War II, many countries, especially the United States, Canada, Australia, and New Zealand, saw a marked increase in the number of babies borne. This "baby boom" generation has been widely researched and continues to have an impact on society and social policy. There are many ways to quantify fertility. For example, birth rate is typically the number of live births per thousand people per year for a given population. The total fertility rate of a population is an estimated measure based on observed age-related fertility rates during a given time period and assuming a woman lives throughout her entire likely reproductive span, or roughly to age 50. It is intended to represent the average number of live births per woman in a given population. However, since human reproduction requires genetic contributions from both males and females and social conventions typically restrict who may reproduce with whom, the male-female ratios in populations can affect actual fertility. Net reproduction rate quantifies the number of daughters borne to a woman, using statistical estimation methods similar to the total fertility rate. This statistic is often used in researching countries that exhibit strong preferences for one sex of child over another or that practice sex selection. Some other possible estimates include gross fertility rate, generational or cohort fertility rate, or completed family size.

In 2010, Russian president Vladimir Putin publicly addressed the growing concern of Russia's declining population, which he attributed to both declining fertility and high death rates, calling it "the most acute problem of contemporary Russia." Sub-replacement fertility rate is a threshold value of the total fertility rate where the number of births is not large enough to replace or maintain a given population at its current level. In theory, each couple must produce two children to replace themselves or, referring to net reproduction rate, each woman must have one daughter to replace herself. In reality, not all people pair and reproduce and early mortality and other factors affect population sizes. Mathematical and statistical models have been used to model average behavior and account for such variables. In the early twenty-first century, the global replacement fertility rate was about 2.3 children per woman: the theoretical value of two, plus a fractional value that adjusts for mortality and other factors. Anything below this value is sub-replacement, leading to a declining overall population. In developed countries, the value was about 2.1 children per woman, while in some developing nations, the replacement rate has been calculated to be as high as 3.3 children per woman. Leslie models, named after population biologist Patrick Leslie, often include fertility matrices based on age groups to model population growth. They are also related to Euler–Lotka equations of population dynamics, named for mathematical demography pioneer Alfred Lotka and mathematician Leonhard Euler.

Fertility Cycles

Individuals seeking to improve their own fertility often rely on various methods to either predict when a woman will be fertile, such as measuring and charting basal body temperature, or to study the viability and motility of male sperm. In the late nineteenth century, physician Mary Putnam Jacobi was among the first to observe biphasic patterns in basal body temperature during menstrual cycles, though the connection with ovulation was not made until the early twentieth century. Studies by many researchers throughout the twentieth century statistically determined patterns in ovulation and fertility, such as the frequency of ovulation, the most probable window of ovulation during the menstrual cycle, and associations between fertility and observable physical characteristics, such as temperature, pain, and mucosal secretions. Many of these studies were the basis for calendar-based methods of fertility planning, such as basal body temperature (BBT) graphs. Beginning in the mid-twentieth century, physicians and others mathematically analyzed and interpreted BBT charts, though some techniques required complete data over long periods, which was considered not to be practical for use by individual couples. In the 1960s, neurologist John Marshall proposed the "three over six" prediction method: a pattern of any three plotted daily temperatures higher than the previous six was a sign of likely ovulation. This method was still in common use at the start of the twenty-first century, though with advances in computing technology, mathematical algorithms for detecting patterns may be used. Alternatively, the Billings method, named for physicians John and Evelyn Billings, is a scoring or quantification system for rating and graphing characteristics of cervical mucus to predict ovulation.

Greater understanding of the biomechanics of conception resulted in new studies of the male role in fertility. Male fertility is often quantified by sperm count or sperm concentration, which is the number of sperm cells per unit fluid volume. The term "oligozoospermia" refers to a sperm count that falls below "normal" as compared to statistically derived reference standards set by the World Health Organization and other agencies. Sperm cells may also be analyzed for abnormal morphology or geometry, which is one of the factors that affects their motility (rate of motion). Mathematical analyses have been used to explore motility. For example, mathematicians David Smith and John Blake created a mathematical model of a swimming sperm cell that they used to explore the fluid dynamic forces between sperm cells and surfaces. Understanding normal sperm motility via such models may help correct motility problems in infertile men and suggest future clinical practices.

Further Reading

Brown, Robert. *Introduction to the Mathematics of Demography*. 3rd ed. Winstead, CT: Actex Publications, 1997.

Poston, Dudley, and Leon Bouvier. *Population and Society: An Introduction to Demography*. Cambridge, England: Cambridge University Press, 2010.

Sarah J. Greenwald
Jill E. Thomley

Fingerprints

Category: Medicine and Health.
Fields of Study: Geometry; Representations.
Summary: Mathematical algorithms help professionals use fingerprints as a means of identification.

The study of fingerprints could be considered both science and art. Fingerprint interpretation and analysis have grown over the twentieth and twenty-first centuries along with the development of new technologies and mathematical tools for imaging processing. Fingerprinting is a recognized method for personal identification and is used worldwide.

A fingerprint is an impression left by the raised portion of the epidermis on the fingers. The epidermal ridges are small corrugations of the skin (with an average of 0.5 mm in breadth) without hair or sebaceous glands but with numerous sweat glands, also found in the toes. The epidermal ridges in a particular area of the inner hands and bottom of the feet have two functions: to provide traction to help people grab objects (the sweat glands moisten the skin, augmenting the security of contact) and to enhance the sense of touch by the stimulation of the underlying nerve. Humans are not the only species with epidermal ridges; some primates, including gorillas and chimpanzees, and koala bears have their own unique prints.

Fingerprint Patterns

There are three general groups of fingerprint patterns: arch, loop, and whorl. They may be divided into subgroups by means of the smaller differences existing between the patterns in the same general group. Fingerprint groups may be also divided into male and female and by age. Historically, the identification of these patterns was done manually in a tedious and time-consuming approach requiring ink, paper, and sufficient knowledge and training of the fingerprint examiner. In the early twenty-first century, automatic fingerprint identification systems can quickly verify a person's identity by searching millions of records in a matter of seconds. Advanced mathematical algorithms are used in forensic science and other areas such as biometric identification, the science of identifying a person using some unique physical characteristic. Correlation-based methods rely on identifying characteristics of print patterns and positioning those characteristics within the pattern, using what are called "registration points." Another mathematically interesting problem is to reconstruct a fingerprint from a partial print or a blurred print.

Other Advances

Other methods to identify humans are used in addition to fingerprints: biometric technology voiceprint, retina/iris scan, hand geometry, and facial recognition. However, fingerprinting is the easiest to use and it provides an average accuracy of 98%.

Wavelets have become an important mathematical tool for fingerprint recognition. This method could be an efficient solution for fingerprint recognition systems because it eliminates the necessity of preprocessing the images, reducing the time required for analysis.

Fingerprint identifications play a vital role in many criminal investigations but there are still challenges, such as identifying the body of a victim of a fire with parts of the fingers burned. Mathematical equations and operators have been used for the calculation of fingerprint probabilities based on individual characteristics, such as only a partial print. The use of digital fingerprints requires more work in description and analysis to avoid ambiguities in identification, such as wrongly convicting an innocent person to prison.

Further Reading

Federal Bureau of Investigation. "The Science of Fingerprints (Classification and Uses)." http://ebooks.ebookmall.com/title/science-of-fingerprints-classification-and-uses-hoover-ebooks.htm.

Hawthorne, M. R. *Fingerprints: Analysis and Understanding.* Boca Raton, FL: CRC Press, 2009.

Komarinski, P. *Automated Fingerprint Identification Systems (AFIS).* London: Elsevier, 2005.

Orton, William. "The Mathematics of Fingerprints." *School Science and Mathematics* 88, no. 1 (1988).

Maria Elizete Kunkel
Maria Elizabeth S. Rodrigues

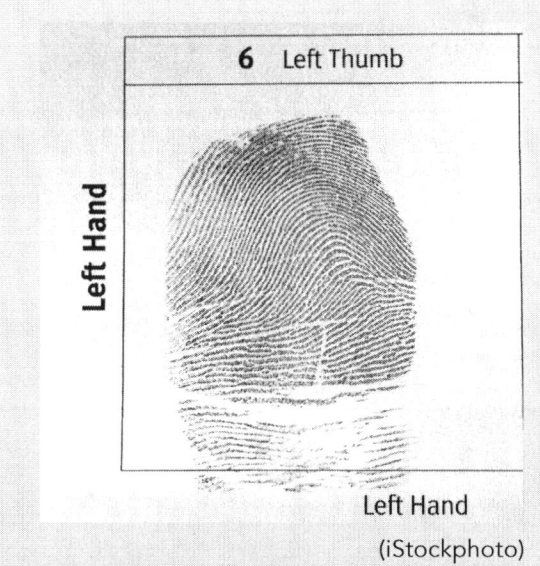

(iStockphoto)

Brief History of Fingerprinting

In 1788, a German scientist, J. C. A. Mayer, presented the theory that each person produces a unique fingerprint. Almost 50 years later, Johannes Purkinje explained that the fingerprints could be classified in patterns that could be recognized. It was the beginning of fingerprints being used to identify individuals. In 1892, anthropologist Sir Francis Galton, cousin of Charles Darwin, published that fingerprints remain unchanged for a person's life and they are permanent. This led to the official use of fingerprints for criminal identification at Scotland Yard. In the early twenty-first century, fingerprint verification has been used as one of the most reliable personal identification methods for criminal investigation or to access control applications.

Firearms

Category: Government, Politics, and History.
Fields of Study: Algebra; Measurement; Number and Operations.
Summary: Mathematicians have long studied and analyzed firearms and projectile motion to create more accurate weapons.

The successful construction and use of many types of offensive weaponry rely on mathematical principles. Ancient people typically used body-powered projectiles, like spears and stones thrown from slings, which required judgments of force and angles to achieve the correct parabolic motion to hit the target. Archimedes of Syracuse designed weapons like mechanical catapults to defend Syracuse from attack by the Romans. The development of gunpowder-propelled field artillery, the successor to mechanical missile weapons like catapults and ballistae, created a demand for sophisticated mathematics. Mathematicians studied and solved problems of ballistic velocities and trajectories to increase accuracy and impact. Handheld firearms of all types rely on similar principles. There are a number of interesting

A collection of World War II firearms. Groups of mathematicians were employed by the war departments both in World War I and World War II for tasks such as creating tables of trajectories for army artillery units. (iStockphoto)

mathematical properties related to firearms, including weapon caliber, rate of fire, rifling, muzzle velocity, and propulsion, as well as telescopic sights and other characteristics. Mathematics training or degrees are suggested for firearms identification and bullet matching, which are increasingly used to match weapons to crimes, and mathematics skills are one of the requirements cited for careers in firearms repair.

Brief History of Firearms

As artillery and projectiles began to play a much larger role in warfare, kings, generals, and powerful concerns in society began looking for more powerful and more accurate weapons. They called upon scientists and mathematicians to address the problem. Niccolo Tartaglia, Galileo Galilei, Evangelista Torricelli, Rene Descartes, Isaac Newton, and Johann Bernoulli are some of the people who worked on the problem of projectile trajectories. Two of the foremost mathematicians to work in this area were Benjamin Robins and Leonhard Euler.

Tartaglia published an important work on cannon trajectory in the sixteenth century. Using the science and mathematics of the time (Aristotelian dynamics, named for Aristotle, and Euclidean geometry, named for Euclid of Alexandria), he thought of the flight of a cannonball as moving from a line with slope determined by the angle of the cannon, the final trajectory by a vertical line and a circular segment on which the apex of the trajectory occurs joining these two lines. In his 1537 text *La nova scientia* and his 1546 text *Questi et inventioni diverse*, he indicated that this was only an approximation to the actual trajectory. However, it was such a good approximation—and so comparatively easy—that it was used by artillery groups well into the eighteenth century. His model took into account the practical knowledge gained through working with gunners and their experience in the field.

Galileo stated that if there is no air resistance, the trajectory of a projectile is a parabola. This conjecture appears first in the work of his student Bonaventura Cavalieri in 1632 and, later, in Galileo's 1638 work *Dis-*

corsi e dimostrazioni matematiche: intorno a due nuoue scienze. Torricelli also worked with Galileo. His book *De motu* contained a geometric method for computing the range of a projectile. Galileo asserted that the path of the trajectory and the shape of a hanging curve (the catenary) are the same, leading him to work with the idea that the trajectory curve is symmetric. This idea results in erroneous computations for range.

In the next era, the important work of Christiaan Huygens, Bernoulli, and Newton on air resistance set the stage for great strides forward in understanding projectile trajectories. The first to explicitly consider air resistance was Benjamin Robins, an English mathematician who was a student of Henry Pemberton and a protégé of Newton. Robins became interested in military engineering in the 1730s from his work on Newton's fluxions and their utility in describing objects in motion. In 1736, he wrote a detailed critique of Euler's *Treatise on Motion* and his extensive use of algebra versus geometry. He was subsequently barred from an appointment as mathematics professor at the new Royal Military Academy in Woolrich in 1741 because of a political dispute. In order to bolster his application for this position, he returned to his work on ballistics and in 1742 published *New Principles of Gunnery*. In 1747, the Royal Society awarded him its prestigious Copley Medal for his work in ballistics. A major contribution of Robins was in determining that the important consideration for ballistics was the initial velocity of the projectile and the effect of air resistance, not the range, which was a function of initial velocity. Experiments showed that the assumption of Huygens and Newton that air resistance was proportional to the square of the velocity was true only at low velocities. Also, Robins hypothesized that lateral deviations were caused by random spinning of the projectile. He advocated the use of rifled barrels with ovoid (rather that spherical) bullets to control this effect.

Robins's work ultimately had a broad impact. In a time of very poor English-Continental mathematical relations, Euler himself found Robins' work from 1742 so important that he translated it into German in 1745 and made extensive additions. He contributed and acclaimed the work of Robins. Napoleon Bonaparte, an avid student of mathematics who is widely considered to have revolutionized the use of field artillery, had Euler's translation translated into French for his study. Euler is credited with bringing the study of trajectory motion into the modern mathematics realm. In his 1753 work, he described motion in terms of second-order differential equations, allowing him to make appropriate changes in assumptions about air resistance and to give better approximate solutions that matched experimental results. Work was undertaken to create tables of trajectories for army artillery units. As technology advanced, mathematics had to evolve to keep pace. There were groups of mathematicians who worked for the war departments in both World War I and World War II. For example, British mathematician John Littlewood improved and simplified calculation formulas for range, flight time, and angle of descent of projectiles and updated ballistics tables. The Applied Mathematics Panel in the United States in World War II looked at various trajectory issues, including aerial dogfights and projectile trajectory. The U.S. Navy maintained the Aberdeen Proving Grounds after the war and had panels of mathematicians there to help model projectile motion and explosions.

There are a number of other interesting mathematical connections related to artillery and firearms, such as caliber and barrel rifling. The caliber of a firearm is the approximate diameter of the barrel and the projectile used in it, usually measured in inches or millimeters. Rifling is traditionally the process of making helical grooves down the entire length of a firearm's barrel to impart a spin to the projectile. Polygonal rifling is another method that shapes the interior of the barrel like a polygon with rounded edges to achieve a similar effect, most commonly with hexagons but sometimes with octagons or decagons. Overall, rifling gives the projectile gyroscopic stability and improves its trajectory. Since a rifled barrel is noncircular, as opposed to a smoothbore (nonrifled) weapon, there are different ways of measuring caliber. In the case of helical rifling, measurements may be taken of the bore diameter, which is the diameter across the lands or high points in the rifling, or the groove diameter, which is the diameter across the grooves or low points. Rifling grooves create striations on the bullet, which, together with caliber, are used in forensics to identify the firearm that shot a bullet. Twist rate for rifling is the distance the projectile must travel down the barrel to complete one full revolution about its own axis, which is often given in units of turns per inches or centimeters. A shorter distance indicates a higher turning rate and a faster spin.

Further Reading

Barnett, Janet Heine. "Mathematics Goes Ballistic: Benjamin Robins, Leonhard Euler and the Mathematical Education of Military Engineers." *BSHM Bulletin: Journal of the British Society for the History of Mathematics* 24, no. 2 (2009): http://dx.doi.org/10.1080/17498430902820887.

McCleary, J., and D. E. Rowe. "Airborne Weapons Accuracy: Topologists and the Applied Mathematics Panel." *The Mathematical Intelligencer* 28, no. 4 (2006).

McMurran, Shawn, and V. Frederick Rickey. "The Impact of Ballistics on Mathematics." http://www.math.usma.edu/people/rickey/talks/08-10-25-Ballistics-ARL/08-10-23-BallisticsARL-pulished.pdf.

2009 Product Engineering Processes. "Archimedes Death Ray: Idea Feasibility Testing." October 2005. http://web.mit.edu/2.009/www/experiments/deathray/10_ArchimedesResult.html.

David C. Royster

Floods

Category: Weather, Nature, and Environment.
Fields of Study: Algebra; Data Analysis and Probability; Measurement.
Summary: Engineering has always been engaged with flood protection and the containment of floodwaters; mathematics is also used to predict flooding.

Although some floods occur with little to no warning, overall patterns of flooding along rivers or streams can be determined based on measurement and the statistical analysis and extrapolation of gathered data, such as a river's historical and current discharge, stage, and flood-stage levels. The resulting data can be used to find the probability of future flooding. Mathematicians and engineers are actively engaged in developing systems to model, predict, and control floods, especially for low-lying areas of the world like the Netherlands and the Mississippi Valley in the United States. Flood prediction and flood control are vital because of floods' potentially devastating impacts—floods are among the leading natural disasters in terms of loss of life and property damage.

Flood Prediction

One of the first steps in flood prediction is the measurement of a river's discharge, stage, and flood stage. The size and flow of rivers are measured using a variety of different methods. Key determinations include the discharge or flow, which measures the volume of water passing through a section of the river in a particular time frame, such as cubic feet per second; the stage, or water surface level over a set criteria, such as sea level; and the flood stage, when a river's overflow will result in widespread inundation or heavy impacts on life and property. Determination of the area of inundation during a flood stage must also take into consideration the topography of the nearby area, such as its slope. During a particular flood, analysts also determine the peak or crest, when the river reaches its highest stage.

Scientists then create flood forecasts based on calculations determined from the statistical analysis of the gathered data. The mathematical calculation of the relationship between an area's precipitation levels and the discharge of nearby rivers and streams relies on a number of complex factors. Geographical factors can include the topography; types of bedrock, soil, and vegetation; and area of the drainage basin. Meteorological factors can include the intensity and duration of precipitation on average, as well as before and during a particular storm. Because of the complexity of the data, forecasters rely on calculating probability based on historical data of peak discharge frequency.

Statistical analysis of the probability of exceeding the average annual peak discharge in a specified time frame can be made for drainage basins for which a series of records of maximum annual discharges (peak flow) are available and ranked from largest to smallest. The calculated probabilities include the probability that a peak flow will be equaled or exceeded within one year, known as the "exceedence probability" and expressed as a decimal fraction; and the recurrence interval, which is the average number of years between past events. The recurrence interval can also be defined as the number of years in which analysts expect a one-time flow that will equal or exceed a peak flow.

The recurrence interval for a particular location can be used to determine the probability of a flood at that location, expressed by the formula

$$P = \frac{1}{T}$$

An aerial photograph of flooding in Port-au-Prince, Haiti, caused by Hurricane Ike in 2008. The U.S. Navy photographed the flood while providing disaster relief support to Haiti. (U.S. Navy, Gina Wollman)

where P is the probability of a flood and T is the recurrence interval. For example, a 100-year recurrence interval would produce a 1% probability of a flood of equal or greater magnitude in a given year. Engineers, scientists, forecasters, and the public must be aware, however, that the resulting probability is an average. For example, a 100-year flood is not statistically expected to occur exactly once every 100 years and two such floods may occur in close proximity.

Graphing and Modeling Floods

Analysts use these statistics in the construction of graphs and tables known as "frequency distributions," which show the probability of various discharges for particular locations and thus the probability of a flood in a particular area. Analysts can utilize a variety of mathematical equations to carry out the statistical analysis needed to create frequency distributions. The most common equations include Normal Distribution, Log-Normal Distribution, Gumbel Distribution, and Log-Pearson Type III Distribution.

Different mathematical methods are used to determine frequency distributions in those locations where recorded data of discharge is unavailable or incomplete. In some cases, analysts use flood frequency estimates from nearby or similar areas with complete data to create estimates for areas that lack data. One commonly used method is the rational method, which utilizes the relationship between peak discharge and the product of drainage basin area, precipitation intensity level, and a standard coefficient based on the drainage basin's land use or ground cover. Other methods allow for the incorporation of changes in a river's discharge over time as well as its peak discharge. The increasing availability of flood-modeling software allows analysts to input data into computers, which then produce flood probabilities and frequency distributions as well as the effects of environmental impacts, such as deforestation and global climate pattern changes, on future flood patterns.

Applications of Flood Models

Meteorologists use flood probabilities and frequency distributions to aid in the issuance of flood watches and warnings. Engineers use flood probability estimates of both magnitude and frequency when constructing and managing flood control structures, such as dams and levees, as well as nearby structures, such as roads and bridges.

The information is also useful when planning to divert or change the course of rivers or streams that frequently flood, increase the slope of the surrounding topography to lessen inundation, create floodway channels, or determine when to lower dam reservoir levels. Governments and other groups use flood probabilities and frequency distributions when planning the location of residences, towns, and industries along rivers and streams.

Further Reading

Baker, Victor R., and R. Craig Kochel. *Flood Geomorphology*. Hoboken, NJ: Wiley, 1988.

Bedient, Philip B., and Wayne C. Huber. *Hydrology and Floodplain Analysis*. Upper Saddle River, NJ: Prentice Hall, 2002.

Bhaskar, Nageshwar Rao. *Regionalization of Flood Data Using Probability Distributions and Their Parameters*. Lexington: University of Kentucky, Water Resources Research Institute, 1989.

Miller, E. Willard, and Ruby M. Miller. *Natural Disasters: Floods:* Santa Barbara, CA: ABC-CLIO, 2000.

Purseglove, Jeremy. *Taming the Flood: A History and Natural History of Rivers and Wetlands*. New York: Oxford University Press, 1988.

Marcella Bush Trevino

Forest Fires

Category: Weather, Nature, and Environment.
Fields of Study: Algebra; Data Analysis and Probability; Geometry; Measurement.
Summary: The spread of forest fires has been modeled for decades to guide firefighting decisions.

Forest service officials have often used controlled burns to reduce the risk of fires spreading by burning the dry vegetation that builds up on the forest floor. Predicting the spread of a fire, whether a controlled burn or a wildfire burning out of control, is of great interest in forest management. Mathematical models take into account various parameters, indices, and activity levels.

Brief History of Forest Fire Modeling

According to Forest Service (a branch of the U.S. Department of Agriculture) documents, the first mathematical model of the spread of fires was developed in 1946 by W. R. Fons. Fons's model was based on approximating the spread as a series of ignitions, with the key elements being ignition time and the distance between particles. Over the years, fire models became more sophisticated, using increasingly complicated mathematical equations, as in Richard C. Rothermel's 1972 differential and integral equation model of fire spread.

With the development of high-speed computers in the last quarter of the twentieth century, simulation models that use large numbers of relatively simple probabilistic and geometric relationships have become more common. In these models, forest fires are represented by a grid of trees where a variety of parameters are set for each tree.

Examples of Forest Fire Models

A very simple simulation of a forest fire can be modeled with a grid of evenly spaced trees and a number cube. Set a forest dryness factor—a set of numbers that, when rolled on a six-sided die, indicate that a tree will catch fire if one of its four neighbors is on fire. For example, a dry forest might be represented by the numbers 1, 2, 3, and 4. In this example, 4/6 or 2/3 of the time, the fire would spread to neighbor trees.

To see how such a simple model works, set the tree in position (3,2) on fire in the grid below and then roll the number cube for the trees in positions (3,1), (2,2),

(3,3) and (4,2) to see if they will catch fire as the original tree "burns out."

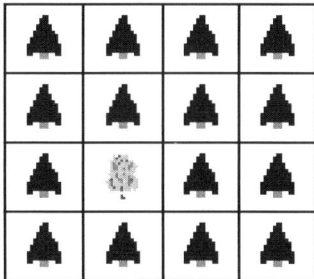

Suppose the number cube rolls are 5, 2, 3, and 2, respectively; then the fire would spread as pictured, with the tree in position (3, 1) remaining unlit.

To complete the simulation, continue rolling the number cube to see if the trees adjacent to the three now on fire will burn. As with all models involving probabilities, it is important to run the simulation a large number of times and to look at the average or most common results rather than to rely on one run of the simulation, such as the forest fire simulation screen shot below:

The simulations in use today to model forest fires are very sophisticated. They include thousands of trees and hundreds of parameters, such as tree size, distribution, and dryness; wind speed and direction; humidity and ambient air temperature; leaf litter buildup; heating, ignition, and burn time; and the geometry of the terrain. These parameters contribute to the calculation of the probability that a tree will catch fire when its neighbors are on fire. Computer visualization software of the early twenty-first century allows programmers to build sophisticated user interfaces for these models in which the spread of the fire can be watched on screen and users can interact with the model, clearing a firebreak or starting a backfire.

Applications of Forest Fire Models

These models can be used to predict how a hypothetical fire might behave or to determine the best intervention in an existing fire, provided the parameter values used in the model accurately reflect the real conditions in the forest. Estimating these parameters poses a challenge to forestry officials—terrain and tree size and distribution are constant in a given forest at a specific time but other parameters, such as tree dryness, wind speed and direction, humidity, and ambient air temperature, vary over time, sometimes significantly.

Failure to accurately gauge parameters in a model can lead to disastrous results. In 2000, the National Park Service developed a fire plan for a controlled burn at the Bandolier National Monument in New Mexico. Now known as the Cerro Grande fire, the wind shifted and strengthened unpredictably, causing the fire to rage out of control, damaging more than 200 homes and 48,000 acres of land in and around the town of Los Alamos.

Agencies and firefighters use a wide variety of National Fire-Danger Rating System (NFDRS) indices and activity levels to monitor and make decisions about fires. For example, the Occurrence Index predicts the potential fire incidence within a rated area. Fire behavior researchers, like George Byram, defined many quantitative measures of fire behavior, such as the definition for fire intensity as the rate of heat energy release per unit time per unit length of fire front, which is defined independently of the depth or width of the fire. The Burning Index (BI) is commonly used to indicate the amount of effort that is needed to contain a given fire. The BI is calculated based on the material that is burning and other factors, including a modification of an equation defined by Byram for flame length. Some people have criticized agencies for failure to use historic data in making future predictions of wildfire hazards, such as recent burn areas in which wildfire is rarely likely to spread.

Newer mathematical models may improve fire forecasts and replace indices like the BI. Statistician Frederic Schoenberg collected and analyzed historic wildfire data in order to build statistical models that clarify relationships such as the apparent linear association between wildfire hazard and average temperature for those that fall below 21 degrees Celsius. While drought is a demonstrated predictor of fires, climatology statistician Sam Shen, atmospheric physicist Robert Field, and earth scientist Guido van der Werf also linked fires in Indonesia with changes in land use and population density. These types of studies have led to quips that only mathematics can prevent forest fires.

Further Reading

Cohen, Jack, and John Deeming. *The National Fire Danger Rating System: Basic Equations*. Berkeley, CA: Pacific Southwest Forest and Range Experiment Station, 1985.

Johnson, Edward, and Kiyoko Miyanishi. *Forest Fires: Behavior and Ecological Effects*. San Diego, CA: Academic Press, 2001.

National Science Foundation. "NSF Discoveries—Improving Fire Forecasts." http://www.nsf.gov/discoveries/disc_summ.jsp?cntn_id=100272&org=NSF.

Rothermel, Richard C. *A Mathematical Model for Predicting Fire Spread in Wildland Fuels*. Report from the Intermountain Forest and Range Experiment Station, Forest Service, U.S. Department of Agriculture, 1972. http://www.treesearch.fs.fed.us/pubs/32533.

Shodor Education Foundation, Inc. "Interactivate: Fire!" http://www.shodor.org/interactivate/activities/Fire.

HOLLY HIRST

Game Theory

Category: Games, Sport, and Recreation.
Fields of Study: Algebra; Number and Operations.
Summary: Game theory models various real-world and hypothetical situations as "games," the play and strategy of which can be analyzed mathematically.

Game theory is the branch of mathematics dedicated to analyzing strategic behavior in different situations. It attempts to describe situations in which several people or entities must make choices even when the outcomes of their decisions rely on the choices made by others. While game theory can be used to address situations typically thought of as games, such as checkers and poker, it can also be used to study situations that are extremely practical and important, such as strategies to use in military operations or auctions and the evolution of species. As with many areas of mathematical modeling, approaching a problem in game theory first typically involves quantifying the objectives and options in terms of algebraic equations and then finding the choice that gives the highest possibility of maximal success.

History of Game Theory

People have studied games and strategies for centuries, but game theory came into its own as a branch of applied mathematics when John von Neumann proved what is known as the "minimax theorem" in 1928. This theorem considers games played between two players in which each player chooses one of a finite number of options, and—depending on the choice made by each player—one of the players gives a certain amount of money to the other player. This is commonly referred to as a "zero-sum game," as the losses incurred by one player exactly equal the gains won by the other player. Von Neumann was able to prove that there is a unique strategy that will maximize a player's winnings (or minimize losings), and one can find this strategy by considering the worst-case outcomes of each of the player's choices and choosing the best-possible, worst-possible outcome. In particular, one would typically like to choose an option that leaves the player indifferent to the choice made by his or her opponent. This work was later expanded by von Neumann and Oskar Morgenstern in their book *Theory of Games and Economic Behavior*, which introduced game theory as a valuable tool for economists.

Rock-Paper-Scissors

An example of the type of strategy that von Neumann wrote about comes up when playing the children's game of rock-paper-scissors. In this game, each of two players chooses one of three possible options (rock, paper, or scissors), and—depending on the choice made by each player—one of the two is declared the winner. In particular, rock beats scissors, scissors beats paper, and paper beats rock. If the two players make the same choice, the game is declared a tie. No matter which choice an opponent makes, one of a player's three options will result in a win, one will result in a loss, and one will result in a tie. Therefore, if the player does not have any inside knowledge of what the opponent will choose, the player will do best by choosing one of the three options at random, each with a probability of one-third.

The Prisoner's Dilemma

The most famous problem in game theory is the Prisoner's Dilemma. The Prisoner's Dilemma is a non-zero-sum game in which there are two participants, each choosing one of two possible outcomes. It is most often described by the following type of story: two criminals, Alice and Bob, are arrested after committing a crime. The police isolate the two prisoners and interrogate them separately. Each criminal must choose whether to confess or to deny the crime, without communicating with the other prisoner. If both confess, they will each get three years in jail. If both deny the crime, there will not be enough evidence to convict them of the felony, but both will get one year in jail. If Alice confesses and Bob denies the crime, then Alice will go free and Bob will go to jail for five years, but if Bob confesses and Alice denies the crime, then Bob will go free and Alice will go to jail for five years. One can see that no matter what Alice chooses to do, Bob will be better off confessing and no matter what Bob chooses to do, Alice will be better off confessing. Because they cannot communicate, one is led to suspect that they will both end up confessing, even though they would both be better off if they both chose to deny the crime. This situation's key principle is how much the criminals trust their partner to deny the crime, rather than do what is in their own self-interest. While this story may seem contrived, it turns out to have many applications in areas such as economics, biology, and political science.

Applications of Game Theory

Much of the research on the Prisoner's Dilemma, as well as other areas of game theory, has taken place at the RAND Institute, a nonprofit think tank originally set up by the United States Army and the Douglas Aircraft Company with a mission "to help improve policy making through research and analysis." Along with

John Forbes Nash, Jr.

John Forbes Nash, Jr., is a mathematician who worked extensively in game theory, in addition to work in algebraic geometry and topology. His best-known work involved finding solutions to games that are not zero-sum games so that players can collectively get better outcomes if they work together than they will get if they work against one another. Nash was born in Bluefield, West Virginia, in 1928 and received his undergraduate degree from Carnegie Mellon University. His dissertation, completed in 1950 at Princeton University, defined the concept that has become known as "Nash Equilibria," which are pairs of choices that two players can make in which neither player is tempted to change their choice. His theory was that most games will eventually work their way to such a situation if they are played repeatedly.

This work, along with subsequent work in this area, led to Nash's being awarded the Nobel Prize for Economics in 1994. In addition to being a mathematician, Nash was a schizophrenic and has spent much of his life dealing with treatments for paranoid schizophrenia, including several prolonged stays in mental hospitals. His life story is the subject of the book and film *A Beautiful Mind*.

then defense secretary Robert McNamara, they developed the game theoretic concept of mutually assured destruction (MAD), which leads to a military doctrine of nuclear deterrence. The idea is that if one country launches a nuclear attack on another, then the conflict quickly escalates until the whole planet is destroyed, and, therefore, such an attack will never take place. This concept has been critiqued by many scholars, but is still an influence on foreign relations today.

While most games in the real world deal with situations in which the players do not have full information or in which there is an element of chance, there is also a strong mathematical study of perfect information games such as checkers and Go. One famous example of such a game is Nim, a game played between two players starting with a number of objects in different piles. On each player's turn, they can remove any number of objects from a single pile. The players alternate turns, and the player to remove the final object loses. This game has been extensively studied and written about by game theorists, such as Elwyn Berlekamp and John H. Conway. It turns out that one of the two players is guaranteed to have a winning strategy, but which player it is depends on the number of piles and the number of objects.

A Game of Pistols

Three people have decided to settle a conflict by firing at each other with pistols. Mr. Pink has a one-third chance of succeeding and killing his opponent, while Mr. Blue has a two-thirds chance, and Mr. Orange is guaranteed to succeed. To even this score, the men will take turns with Mr. Pink taking first shot, followed by Mr. Blue, and then Mr. Orange. A natural question in this situation is whether Mr. Pink should shoot at Mr. Blue or Mr. Orange first. It turns out that the answer is neither. Using game theory, one can show that Mr. Pink has a better chance of surviving if he shoots into the air and intentionally misses both of the other players.

Further Reading

Berlekamp, Elwyn, John H. Conway, and Richard Guy. *Winning Ways for Your Mathematical Plays*. 2nd ed. Natick, MA: A K Peters, Ltd., 2001.

Nash, John. "Equilibrium Points in *N*-Person Games." *Proceedings of the National Academy of Sciences of the United States of America* 36, no. 1 (1950).

von Neumann, John, and Oskar Morgenstern. *Theory of Games and Economic Behavior*. 4th ed. Princeton, NJ: Princeton University Press, 2007.

Poundstone, William. *Prisoner's Dilemma: John von Neumann, Game Theory and the Puzzle of the Bomb*. New York: Anchor, 1992.

Darren Glass

Genetics

Category: Medicine and Health.
Fields of Study: Data Analysis and Probability; Number and Operations.
Summary: Bioinformatics and probability theory come into play in the study of genetics.

Issues related to genetics are no longer exclusively discussed in academic circles. The lay community every day accesses a large amount of information through the mass communication vehicles that enable the socialization of knowledge related to heredity and biotechnology. Paternity tests, transgenic plants, early diagnoses in medicine, gene therapy, and cloning are no longer exclusive subjects of specialized research centers and can be easily researched in the media and found in movies, cartoons, and on the Internet. These are examples of how closely aligned this area of science is to modern society and how broad the possibilities are for development. Mathematical tools are essential for the analysis and interpretation of data related to these genetic processes that otherwise would become empty of real meaning. From the simple knowledge of probability to the most powerful algorithms associated with genetic engineering techniques, probability is necessary to elucidate the most difficult questions surrounding the genetics field. Sharon Grossman is one mathematician who has notably contributed to genetics research through her investigations of gene groupings based on geographic location.

Genetics is the field of biology that studies the chemical nature of hereditary material and the mechanism responsible to transfer information contained in genes. In general, reproduction is constituted by a series of events that result in a randomized combination of gametes. This process involves the mixing of thousands of information packets and results in the production of a new living being.

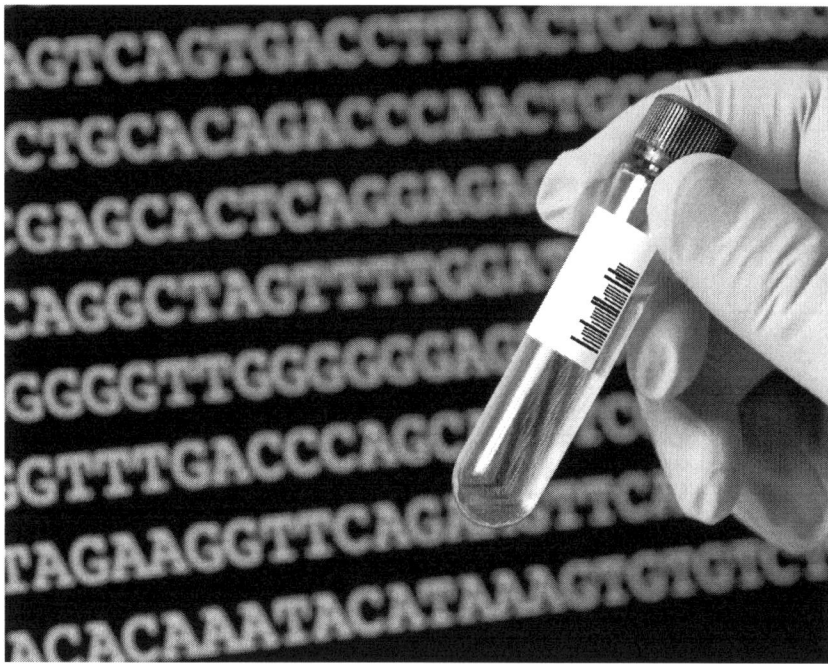

Powerful computers can analyize a patient's DNA data, and have revealed that most of the DNA molecule is not involved in protein-coding. (iStockphoto)

Early Findings

The first steps of genetics were performed by Austrian Gregor Mendel (1822–1884), who, for many years, crossed varieties of peas. After obtaining numerous generations of these plants, he observed differences in the types of progeny formed and identified the proportion of each of these features in future generations. His main findings showed that specific factors were transmitted by parents to offspring. He also found that these factors occur in pairs and that their descendants receive one from each parent. Crosses made with peas (called "hybrids") had particular characteristics, like seed color. By calculating their frequencies, Mendel realized that the prevalence of these factors was different in several generations. Some manifest themselves only when appearing in double dose (recessive), while others in a single dose determined the characteristic (dominant). These findings served as the basis for developing laws on inheritance, which came to be called the "first and second laws of Mendel."

Genetic Probability

Probabilities are used to express the chance of occurrence of an event. They represents a possibility, not a conviction. The probabilities can be expressed in several ways, including fractions, percentages, and decimals. For example, the chance of occurrence of a biological event can be expressed as "50%," "0.50," or "1/2."

Many genetics calculations are solved using probability. Mendel used mathematical rules previously used for common events, such as a coin toss (individual events), or combined events, such as the simultaneous release of multiple dice.

Genotype is the set of genes from one living being, the frequency of these genes, and can be calculated mathematically. The calculations performed in the theory of probability do not determine the appearance of a particular genotype—they merely represent the chance this event will occur. In practical terms, genetic calculations allow one to determine the probability, for example, of two individuals with dark eyes to conceive a child with blue eyes (a recessive gene). This event is possible if both parents are hybrids, in which case the probability in each pregnancy is 25%.

The application of rigorous scientific method and careful statistical research of some characteristics led Mendel to conclusions that still underlie modern genetics in the early twenty-first century. Not until in 1900 could the work of three independent researchers—Hugo de Vries, Karl Correns, and Erich Tschermak—show that Mendel's conclusions were correct.

Modern Genetics Research

For a long time it was believed that protein was the molecule that contained genetic information. Biochemical studies allowed the identification of a molecule able to replicate and thereby allow a flow of identification information: deoxyribonucleic acid (DNA)—the molecule associated with heredity. In 1953, James Watson and Francis Crick published in *Nature* the model of the DNA molecule. Understanding the complex spatial geometry of DNA allowed researchers in the early 1960s to prove that the code was formed by groups of three nucleotides that were repeated in complementary sequences. It was noticed that sequences of the DNA molecule were able to be expressed as proteins with the participation of another nucleic acid: ribonucleic acid (RNA). Studies of the most primitive life forms, like bacteria, also led to knowledge related to peculiarities of DNA activity, as well as its transmission and their biochemical behavior.

The challenge became the elucidation of the genome (the entire set of genetic information that is found in the chromosomes) from a living organism. After advancing with some simple life forms, such as bacteria and protozoa, the Human Genome Project (1988–2003) arose. International cooperation efforts were necessary to decipher the sequence of 3 billion base pairs of DNA subunits found in human chromosomes. Powerful computer programs and the use of combinatorial analysis revealed that most of the DNA molecule is not involved in protein-coding. It is now known, however, that the role of this DNA is very significant, especially for matters pertaining to evolution, and it is responsible for many adaptive differences between species.

Genetic Variability

The prevalence of certain genes in a population depends on how the expression of a particular feature is selected by the environment and is related to the presence of other genetic variation factors, such as genetic mutations, numbers of crosses, or natural events that abruptly decrease the frequency of certain genes in a population (for example, earthquakes, fires, or floods).

How is it possible to evaluate this natural dynamic that sometimes takes decades or even centuries to occur? Since a group within the set of genes undergoes a random process of transmission, it cannot be adequately studied without resorting to mathematical tools to assess the frequency of certain genes in a population and the possible consequences of this variability for that group. Wild populations (animals, plants, or, specifically, humans) are subject to phenomena—such as gene recombination, mutation, and gene conversion, which is the change of position of genes within a chromosome—that lead to the emergence of genetic variability. Genetic mathematics aims to understand how genetic changes occur for individuals both within species and over time.

Several phenomena are responsible for genetic variability. Crossing-over, for example, is a phenomenon in which parts of chromosomes are broken and glued in different positions, generating a larger mix of information and expression of phenotypes (physical or physiological), which results in an increased possibility of adapting to the environment in which the individual belongs. Random events observed in gene transfer result in the formation of functional characteristics and patterns that may often cause trouble and injury, but that is partly responsible for the possibility of evolution.

Bioinformatics

Genetics is an area of study that uses the theories of probability and the handling of large volumes of data. The difficulties in performing complex calculations—far more advanced than the calculations made by Mendel—necessitated the use of information technology in studies of biological and genetic research. Bioinformatics is the application of computer systems in the processing of biological and biomedical data. It is an interdisciplinary science that aims to develop and apply computational techniques to study genetics, molecular biology, and biochemistry. Without this tool, it would be impossible to perform thousands of mathematical operations in real time.

In bioinformatics, gene sequences are analyzed and stored in databases, manipulated, and analyzed using specific software. Databases allow scientists to get information from other laboratories and also to share the genetic sequences. Despite the efforts of international collaboration in this area, the patenting of genes clashes science with ethical issues regarding the detention of the natural heritage of knowledge as private property. Laws regarding other issues related to cloning and gene manipulation vary according to country.

Genetic engineering uses principles formulated many years ago. The development of refined methods using molecular biology techniques allowed the manipulation of genetic material, known as "recombinant DNA technology" or "genetic engineering." Once DNA fingerprinting had become associated with the identification of individuals, great hopes arose regarding the possibility of isolating and cloning genes to replace defective genes as therapy.

Further Reading
"Genetics Home Reference: Your Guide to Understanding Genetic Conditions." U.S. National Library of Medicine. http://ghr.nlm.nih.gov.
Lachowicz, M., and J. Miekisz, eds. *From Genetics to Mathematics*. Vol. 79 of *Series on Advances in Mathematics for Applied Sciences*. Hackensack, NJ: World Scientific Publishing Company, 2009.
"Learn. Genetics: Genetics Science Learning Center." University of Utah. http://learn.genetics.utah.edu.

Maria Elizete Kunkel
Maria Elizabeth S. Rodrigues

Geometry of the Universe

Category: Space, Time, and Distance.
Fields of Study: Geometry; Measurements; Representations.
Summary: Characteristics of the universe such as size, shape, and composition have long concerned mathematicians and astronomers and over the course of history various models have been offered.

The shape of the universe and its geometry have been the topic of human interest for millennia. Researchers in scientific disciplines such as physics, astronomy, and cosmology, along with mathematicians, especially those working in geometry, are seeking to discover what shape the universe is, whether it is finite or infinite, and how many dimensions it has. Not only do researchers investigate this topic; it is also popular for philosophical debates in the media and educational settings, for example, as the theme of Mathematics Awareness Month in 2005. Generally speaking, the density of the universe determines its geometry. The shape of the universe could therefore be estimated by measuring the average density of the matter within it, assuming that all matter is evenly distributed—though there might be considered distortions caused by very dense objects with mass accumulated locally, such as galaxies. This assumption is well justified by cosmological observations showing that, while the universe appears to be weakly inhomogeneous and anisotropic locally, on average it is homogeneous and isotropic. Therefore, all considerations about the geometry of the universe have to be seen from two perspectives: the local geometry that is related to the observable universe and the global geometry related to the universe as a whole, where also that is included which humans have yet to be able to measure in the early twenty-first century.

Measurements are closely related to the origins of geometry, a discipline flourishing more than 5000 years ago from the early stages of the human civilization in ancient Egypt and later in ancient Greece, and are practical and necessary in connection to the geodetic measurements of Earth. Later, developed as a theoretical abstract branch of mathematics, geometry offered mathematical background for the description of geometric abstract spaces with more dimensions, which cannot be visualized in the three-dimensional spaces, but can be used as models in modern physical and cosmological theories describing the possible form, structure, and principal laws of the universe.

From the History
For thousands of years, people believed that the universe revolved around Earth, and astronomers created mathematical models to explain observations in the sky. Eudoxus of Cnidus created a model containing rotating spheres centered about Earth. With this model, Aristotle was able to partially explain some of the planetary motions by rotating the spheres at different velocities, but other observations, such as differences in brightness levels, could not be resolved.

In the next century after Aristotle, Euclid of Alexandria expressed the parallel postulate. While it was not linked with models of the universe at the time, it was to eventually take on an important role in the geometry of the universe. Euclid is the author of the famous *Elements*, one of the earliest and most influential works in the history of mathematics, consisting of 13 books. Here, all principles of the geometric space, today called "Euclidean," were deduced in the form of mathematically proved propositions and constructions from a small set of postulates and definitions. Postulates were not proved or demonstrated, but considered to be self-evident and true. They described all basic relations and measures between ideal geometric figures as points, lines, triangles, circles, or solids, and also numbers that were treated geometrically as line segments with various lengths. The introduced list of postulates referred to the following five groups of relations: incidence, congruence, order, continuity, and parallelism.

The fifth postulate about parallelism says: "If a straight line falling on two straight lines makes the interior angles on the same side less than two right angles, the two straight lines, if produced indefinitely, meet on that side on which are the angles lesser than the two right angles." From the time of its publication until the late nineteenth century, this postulate, apparently different from all others and of more complicated form, attracted mathematicians, who strived to prove it as a consequence of the first four groups. New equivalent formulations of this famous parallel postulate appeared. The most familiar form is this: "Through a point not on a given straight line, at most one straight line can be drawn that never meets the given line" (see Figure 1).

All efforts to prove the fifth parallel axiom appeared to be pointless. On the contrary, different possible formulations of this special property were introduced, as the negations of Euclid's postulate, revealing thus the existence of new kinds later called "non-Euclidean" geometries with unusual properties emerging from these formulations.

Figure 1. Axiom on parallelism.

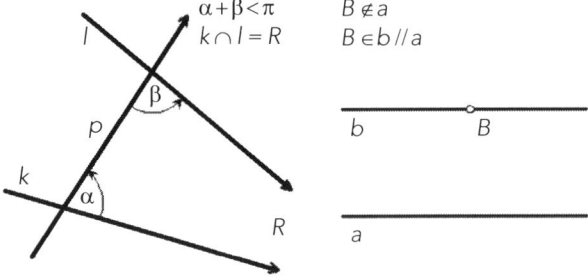

Triangles in Different Geometric Spaces

Among many consequences of the validity of Euclid's postulate on parallelism, the most striking one is about the sum of the triangle interior angles. The measures of the interior angles of a triangle in the Euclidean space always add up to exactly 180 degrees. This property is related to the planar triangles located in a flat plane. Examples of triangles in non-Euclidean geometries are spherical triangles and hyperbolic triangles. Here, the sum of measures of the interior angles of a triangle is always more than 180 degrees in the elliptical non-Euclidean space, while in the hyperbolic non-Euclidean space it is always less than 180 degrees.

Geometric Spaces

Even in the face of overwhelming evidence, it took a long time for humanity to accept that Earth is not at the center of the universe because this revolution required an imaginative leap that surpassed problematic religious and philosophical implications. In his famous work, the *Almagest*, Claudius Ptolemy, a second-century philosopher, refined and improved an Earth-centered model based on the earlier work of Apollonius of Perga and Hipparchus of Rhodes. In the Ptolemaic universe, planets now moved along epicycles, which had circles attached to the spheres around Earth, and yet this model still did not completely resolve contradictions with astronomical observations. Aristarchus had suggested a heliocentric system, and in the sixteenth century, Nicolaus Copernicus gave substance to Aristarchus's ideas by carrying out the detailed mathematical calculations. His model still utilized epicycles in order to explain the circular motion of the planets, but it placed a motionless sun close to the center of the universe. Johannes Kepler revolutionized astronomy by finally overthrowing the stranglehold of purely circular motions. His introduction of elliptical orbits together with his other two laws of planetary motion form the basis of celestial mechanics to this day. They were also critical in the formulation and verification of

Sir Isaac Newton's laws of gravity and of motion, which in turn became the basis for cosmology for the following two centuries.

Around 1830, Hungarian mathematician János Bolyai and Russian mathematician Nikolai Ivanovich Lobachevsky published their papers on non-Euclidean geometry, independently and unaware of each other—hyperbolic geometry is therefore also called Bolyai–Lobachevskian geometry. The famous mathematician Johann Karl Friedrich Gauss explored such geometry about 20 years earlier, but he never published his work. Lobachevsky developed a theory of a new geometric space, in which the fifth postulate was not true, by negating the Euclid's postulate about the existence of a unique parallel to a given line. He stated a new, nowadays called the Lobachevsky, axiom of parallelism: "Through a point not on a given straight line, at least two different lines can be drawn that never meet the given line." Lobachevsky based this reasoning on his own findings received from measuring distances of stars calculated from their trajectories traced on the celestial sphere because of the movements of Earth in the solar system. In his gigantic triangles, the sum of the interior angles measured less than 180 degrees. Bolyai worked out a geometric theory whereby both the Euclidean and the hyperbolic geometry were possible, depending on a special introduced parameter. Bolyai wrote in his work that it is not possible to decide whether the geometry of the physical universe is Euclidean or non-Euclidean through mathematical reasoning alone, and he regarded this to be a task for the physical sciences.

Bernhard Riemann was a German mathematician who founded a new field of geometry, later called the "Riemannian geometry," in his famous lecture in 1854. He constructed an infinite family of non-Euclidean geometries by giving a formula for a family of Riemannian metrics on the unit ball in the Euclidean space. His theory of Riemannian surfaces—which can be divided into three types: hyperbolic, parabolic, and elliptic or spherical corresponding to negative, zero, or positive curvature—can be generalized by his uniformization theorem in terms of conformal geometry. Every connected Riemann surface X admits a unique complete two-dimensional real Riemannian metric with constant Gaussian curvature equal to –1, 0, or 1 inducing the same conformal structure. The surface X is then called "hyperbolic," "parabolic," and "elliptic," respectively, according to its universal cover.

Later on, Riemann's remarkable work was elaborated by German mathematician Felix Christian Klein, who established a new classification of geometric spaces based on algebraic theory of the underlying group of transformations and their invariants, which is known as the "Erlangen program" presented at the University of Erlangen in 1872. Basic properties of a specific geometry can be represented as sets of invariant properties of the space figures under a given group of transformations. This definition of geometric spaces encompassed both Euclidean and non-Euclidean geometry in a unifying theory of geometric spaces, taking into consideration not only geometric figures and the space dimension, but also specified geometric transformations and their invariants.

The development of non-Euclidean geometries was inevitably important to physics in the twentieth century. Modern geometry shows multiple strong bonds with physics, exemplified by the links between Riemannian geometry and relativity. In 1917, Albert Einstein used Bernhard Riemann's mathematics in order to present a model for the universe that was consistent with his theory of relativity. His model was based on a finite spherical universe. Geometry, where the curvature changes locally from point to point, is the Riemannian geometry of continuous manifolds. One of the youngest physical theories, string theory, is also very geometric in flavor.

Dimensions: Shape of the Universe

There is a direct link between the geometry of the universe and its shape. The homogeneous and isotropic universe allows for a spatial geometry with a constant curvature, and three different possible types of geometric spaces can be distinguished, depending on the sign of the curvature.

If the density of the universe equals exactly the critical density, then the geometry of the universe is flat, like a plane. One has to consider a geometric space with zero curvature and Euclidean geometry as described by Euclid. As Euclid's fifth postulate on parallelism is true, the sum of the triangle's inner angles equals exactly 180 degrees, and light photons traveling on parallel lines never meet each other (see Figure 2).

Figure 2. Planar parabolic Euclidean geometric space.

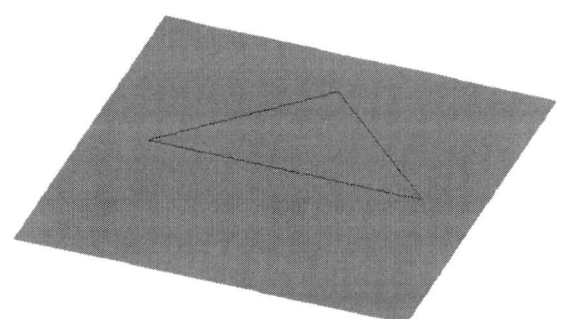

Figure 3. Spherical non-Euclidean geometric space.

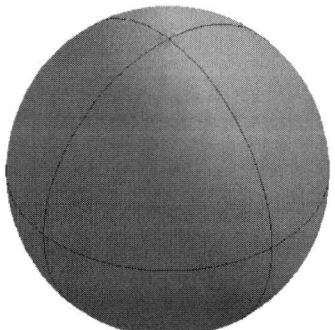

If the density of the universe exceeds the critical density, then the geometry of space is closed and positively curved like the surface of a sphere. No parallel lines exist, and the sum of the triangle's inner angles is more than 180 degrees. This implies that, initially, parallel photon paths converge slowly. Eventually, they cross and return back to their starting point if the universe lasts long enough (see Figure 3).

If the density of the universe is less than the critical density, then the geometry of space is open, negatively curved like the quadratic surface called "hyperbolic paraboloid." Infinitely many parallels exist through a point to a given line and the sum of the triangle's inner angles is less than 180 degrees. Parallel photon paths can be considered as traveling to infinity in different directions from one starting point (see Figure 4).

Global geometry describes the topology of the whole universe—the observable part and beyond. For a flat spatial geometry, any topological property may or may not be directly detectable, as the scale of all such properties is arbitrary. Probability to detect the topology of spherical and hyperbolic geometries by direct observation depends on the spatial curvature. Using the radius of curvature as a scale, a small curvature of the local geometry, with a corresponding scale greater than the observable horizon, makes the topology difficult to detect. In a hyperbolic geometry, the radius scale is unlikely to be within the observable horizon, while a spherical geometry may have a radius of curvature that can be detected.

There are three primary methods to measure curvature: luminosity, scale length, and density. Luminosity requires an observer to fix some standard source of light, such as the brightest quasars, and follow them out to high red shifts. Scale length requires determination and usage of some standard size, which can be the size of the largest galaxies. Density is a number of galaxies in a chosen box as a function of distance. Recently, all these methods have been inconclusive because the size and number of observable galaxies and their brightness are changing with time in unpredictable ways. As of 2011, the cosmological measurements were consistent with the model of a flat universe, based on data from sources such as NASA's Wilkinson Microwave Anisotropy Probe (WMAP). NASA has declared the universe to be flat within a 2% margin of error.

Two following investigations are decisive in the study of the global geometry of the universe:

- Whether the universe is a compact space or it is infinite in extent
- Whether the topology of the universe is simply or nonsimply connected

Figure 4. Hyperbolic non-Euclidean geometric space.

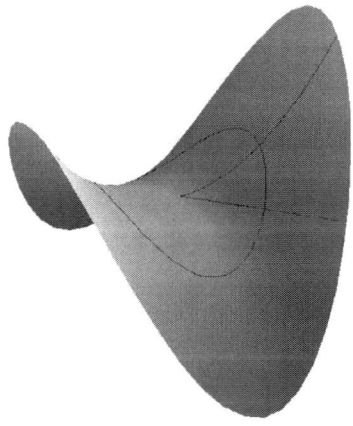

Both of these topological properties depend on the mass distribution and, therefore, on the total strength of gravitation within the universe. However, each of them implies a different history and future development of the universe:

1. If the universe is a space with negative curvature, there is insufficient mass to cause the universe to cease expansion. Therefore, the universe has no boundaries, and it will continue expanding forever, ending in a Heat Death. This model of the universe is presented as an "open universe."
2. If the universe is a space with zero curvature, there is exactly enough mass to stop its expansion, but this will take an infinite amount of time. In this case, the universe has also no bounds and will expand forever; but after an infinite amount of time, the rate of its expansion will be gradually approaching zero. This is a "Euclidean flat universe" model.
3. If the universe is a space with positive curvature, there is more than enough mass to stop its expansion. The universe is not infinite, but it is endless. The present expansion of the universe might eventually stop and turn into a contraction, and the universe will start collapsing on itself. This model is called a "closed universe."

Scientists still do not know which of these three scenarios of the future of the universe could be correct, as they have not yet been able to determine exactly how much mass is in the universe.

If the three-dimensional manifold of a spatial section of the universe is compact, then the universe has a definable volume, as on a sphere. If the geometry of the universe is not compact, then the universe is infinite in extent with no definable volume, such as the Euclidean plane. Therefore, if the spatial geometry is spherical, then its topology is compact, while for a flat or a hyperbolic spatial geometry the topology can be either compact or infinite.

Particle physics, quantum field theory, and cosmological theories led to a revolution in thought and new paradigms of subatomic matter that require the existence of a so-called hyperspace, which is an ultimate universe of many dimensions. In an ongoing quest for a synthesis of quantum mechanics and relativity physics into a superstring theory of universe unifying four fundamental forces (gravity, electromagnetism, and the strong and weak nuclear forces), the idea of a Theory of Everything has been born. This unified field theory, as it is understood in the early twenty-first century, does not preclude any of such hypotheses as, for instance, the existence of superstrings, black holes, wormholes, other parallel universes, and time travel ideas. Modern physics still needs a more powerful mathematical theory and topology of the 10-dimensional space

Dimensions

Dimensionality of the geometric space is an intrinsic characteristic that is understood and perceived differently for the space inhabitants (locally) than from the global point of view of the outside observers. Inhabitants of the three-dimensional space cannot easily realize the fourth dimension, similarly to the behavior and abilities of inhabitants of a two-dimensional space, Flatland. English mathematician and writer Edwin A. Abbott explored the nature of dimensions in his novel *Flatland: A Romance of Many Dimensions* that appeared in 1884, where he predicted the possible existence and reality of the fourth dimension of the universe. Flatlanders are not able to imagine the third dimension existing outside their living environment of the two-dimensional space, which is, however, quite natural for the three-dimensional space inhabitants. His work inspired mathematicians to develop considerations of how higher dimensions could appear to human beings as inhabitants of the universe, provided this can be considered as a three-dimensional surface of a four-dimensional space-time. Many books and films appeared, describing the idea of dimensionality and its perception; for example the short film *Flatland* produced by Seth Caplan in 2007, or the computer animated film *Flatland* directed and animated by Ladd Ehlinger, Jr., in Lightwave three-dimensional software.

in order to understand completely our expanding and evolving cosmos. The theory of hyperspace introduced by American mathematician Michio Kaku may be the leading candidate for the Theory of Everything, for which Albert Einstein spent the last years of his life searching.

When, in 1990, scientists sent the Hubble Space Telescope into space, they did not expect to find that the expansion of the universe was speeding up, nor did they realize the existence of the black matter and the dark energy that became the dominant force in the universe, recently accelerating its expansion. The James Webb Space Telescope, NASA's next orbiting observatory and the successor to the Hubble Space Telescope, is scheduled to be launched in 2014 to distant orbits. This infrared telescope detecting infrared radiation will be capable of seeing wavelengths of light difficult to observe from Earth, thus opening new horizons of the visible universe. It is hard to imagine and predict what discoveries and answers to the mysteries of the universe scientists will gain using its observations in the future.

Further Reading

Abbott, Edwin A. *Flatland: A Romance of Many Dimensions.* Los Angeles: Indo-European Publishing, 2010.

Frank, Adam and Erika Larsen. "Is the Universe Actually Made of Math?" *Discover Magazine* (July 2008).

Kaku, Michio. *Hyperspace: A Scientific Odyssey Through Parallel Universes, Time Warps and the Tenth Dimension.* New York: Oxford University Press, 1994.

Rucker, Rudy. *Spaceland.* New York: Tom Doherty Associates, 2002.

Stewart, Ian. *Flatterland.* New York: Perseus Publishing, 2001.

Weeks, Jeff. *The Shape of Space: How to Visualize Surfaces and Three-Dimensional Manifolds.* New York: Dekker, 1985.

Yau, Shing-Tung and Steve Nadis. *The Shape of Inner Space: String Theory and the Geometry of the Universe's Hidden Dimensions.* New York: Basic Books, 2010.

Daniela Velichová

Geothermal Energy

Category: Weather, Nature, and Environment.
Fields of Study: Algebra; Data Analysis and Probability; Measurement.
Summary: Geothermal energy can be harnessed for domestic heating or to produce electricity via steam turbine.

"Geothermal" refers to heat from the interior of Earth generated from the forces that led to the planet's creation and the ongoing slow radioactive decay that continues to generate thermal activity. While Earth's surface is relatively cool, temperatures increase dramatically with depth, which is known as a region's "geothermal gradient." The interiors of continents tend to have lower gradients than "spreading center" regions, where continental tectonic plates are slowly separating. A prime geothermal area is along the Ring of Fire rimming the Pacific Ocean's eastern, northern, and western coasts.

High geothermal gradients make prime candidates for geothermal energy projects. However, the average gradient is approximately 2.5–3 degrees Celsius per 100 meters. Approximately 6000 kilometers beneath the surface, molten rock reaches temperatures of approximately 5000 degrees Celsius. A small portion of this extreme heat makes its way to the surface as steam through cracks and fissures. Geothermal leakage to the surface leads to dramatic volcanic eruptions as well as to the formation of hot springs and geysers. Geothermal-warmed, mineral-rich waters have long been considered to be sacred or to have healing properties by many people. Geysers such as Old Faithful in Yellowstone continue to attract visitors from around the world.

Mathematicians, geothermal engineers, geologists, and other scientists use mathematical methods to research various aspects of geothermal processes, such as the deformable, porous properties of soil and rock that allow geothermal heat to make its way to the surface. These studies have broad applications in many scientific areas, including the way brains deform during neurosurgery and in industrial injection molding. In other cases, Lagrangian–Eulerian flow models, named for Joseph Lagrange and Leonhard Euler, are used to model characteristics such as precipitation and transport, which have applications for engineering geothermal reservoirs and isolating radioactive waste.

Stochastic models for system optimization and control as well as geometric models also help mathematicians understand geothermal heat. Many are working on computer models to update, integrate, and expand the U.S. Geological Survey's MODFLOW, a three-dimensional finite-difference groundwater flow model first published in 1984 and widely used for research and industrial applications.

Geothermal Heating

As long ago as the nineteenth century, scientists and engineers began to develop geothermal-based applications for chemistry and heating, though there is evidence that even prehistoric people built dwellings around naturally occurring geothermal heat sources. With abundant geothermal resources, Iceland began to emerge by the late 1920s as a world leader in the use of geothermal energy for domestic heating and cooling. Advances since that time have led to the development of geothermal heat pump systems. During cold periods, heat pumps transfer to buildings heat from either the ground (beneath the frost line) or from the bottom of ponds. During warm periods, the process is reversed and heat is taken from buildings and put into the ground or ponds. However, purposeful movement of water on a large scale can have geological consequences. For example, in Venice, the removal of subsurface water resulted in subsidence (settling of loose, porous soil), which lowered some buildings. Adding or subtracting water from one part of a geothermal field can affect all aspects of the field, including system pressure and surface vents. Seismologists use mathematical models describing the behavior of deformable porous rock and soil to predict where events like earthquakes might occur as a result of water-pumping activities.

Geothermal Electricity

Geothermal resources can also be used to produce electricity. The first geothermal electric power plant was built in Larderello, Italy, in 1904. Japan and the United States followed suit in 1910 and 1921, respectively. The spread of geothermal energy has been slow in the decades since. However, because of concerns regarding global warming and a quest to develop nongreenhouse gas (GHG)–emitting energy technologies, geothermal power generation has received more attention.

There are two types of geothermal power plants, both of which rely upon the production of steam to drive the conventional turbines that create electricity. Electricity can be produced directly from steam if the temperatures are at a minimum of 95 degrees Celsius (200 degrees Fahrenheit), and higher outputs are possible after temperatures crest at 175 degrees Celsius (350 degrees Fahrenheit). At the Geysers geothermal power plant in California, steam at a temperature of approximately 235 degrees Celsius (455 degrees Fahrenheit) is used to directly drive turbines. At lower temperatures, geothermal heat can still be used, but it relies upon specialized fluids that have a low boiling point capable of producing high pressures, rather than natural steam.

While the capital costs are high for both types of geothermal electricity, once in production it has several advantages over other forms of electricity generation. Like wind, its fuel costs are negligible. Similar to wind and nuclear power, once constructed, geothermal plants produce far fewer GHG emissions

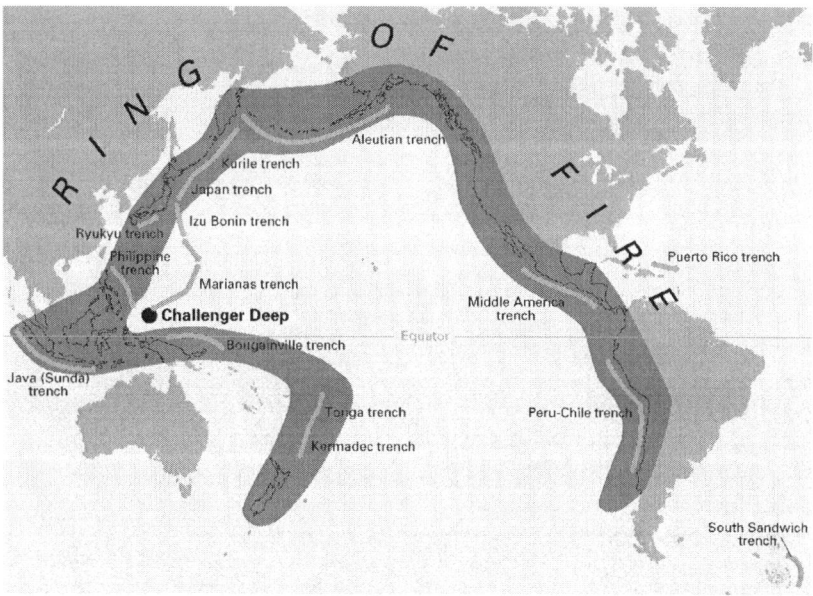

Volcanic arcs and oceanic trenches partly encircling the Pacific Basin form the Ring of Fire, a zone of frequent earthquakes and volcanic eruptions. (U.S. Geological Services)

than traditional fossil fuel plants. Geothermal also has advantages over other alternative energy producers. Unlike wind, which is intermittent because of its dependency on weather conditions, geothermal electricity can be relied upon to produce consistent baseload power. Geothermal plants are also less intrusive visually than large wind farms and tend to draw less public attention.

Geothermal also has two key advantages over nuclear generation. Nuclear power plants are dependent upon a finite resource (uranium), and nuclear waste disposal is both controversial and costly. In contrast, geothermal generation depends on a virtually infinite source (heat generated in Earth's interior), and there are no long-term waste issues.

Popular and government interest in geothermal energy and its advantages over both traditional and alternative electricity generating options led to a 20% increase in global geothermal electricity production between 2005 and 2010. In addition, there has been a 52% increase from 2007 to 2010 in the number of countries developing geothermal resources.

Despite the increasing numbers, geothermal energy production continues to significantly lag behind other electricity sources at the start of the twenty-first century. In part, this lag is the result of a perception that there are a limited number of high-quality geothermal sites that would enable geothermal energy to become a major producer. In addition, there are technical, permitting, and electric transmission issues that drive up capital costs and inhibit substantial expansion.

Further Reading

Bundschuh, Jochen, and Mario Suárez. *Introduction to the Numerical Modeling of Groundwater and Geothermal Systems: Fundamentals of Mass, Energy and Solute Transport in Poroelastic Rocks*. Oxfordshire, England: Taylor & Francis, 2010.

Dickson, Mary H., and Mario Fanelli. "What Is Geothermal Energy?" International Geothermal Association. http://www.geothermal-energy.org/314,what_is_geothermal_energy.html.

"Geothermal Energy: Tapping the Earth's Heat." *National Geographic*. http://environment.national geographic.com/environment/global-warming/geothermal-profile.

Holm, Alison, Leslie Blodgett, Dan Jennejohn, and Karl Gawell. *Geothermal Energy: International Market Update*. Washington, DC: Geothermal Energy Association, May 2010.

Idaho National Laboratory. *The Future of Geothermal Energy: Impact of Enhanced Geothermal Systems (EGS) on the United States in the 21st Century*. Cambridge, MA: MIT Press 2006.

Lerner, K. Lee, and Brenda Wilmoth, eds. "Geothermal Gradient." In *World of Earth Science*. Farmington Hills, MI: Gale Cengage, 2003.

National Energy Board. *Emerging Technologies in Electricity Generation: An Energy Market Assessment*. Ottawa : Her Majesty the Queen in Right of Canada as represented by the National Energy Board, 2006.

U.S. Government Accountability Office. "Renewable Energy: Increased Geothermal Development Will Depend on Overcoming Many Challenges." May 2006.

Jason L. Churchill

Gravity

Category: Space, Time, and Distance.
Fields of Study: Algebra; Measurement.
Summary: Our understanding of gravity has changed considerably over time, such that a history of gravity is virtually a history of physics. Researchers study many different effects and conceptualizations of gravity, some of which are very far from Isaac Newton's falling apple.

On the surface of the Earth, every object has some weight, which is simply the gravitational force that Earth exerts on it. In reality, minuscule gravitational forces are exerted on every atom of every object, the net effect of which is the same as the effect of a single force (the weight) acting at a single point, the center of gravity (CoG). If the object is sitting on a table, the downward force of gravity is balanced by the upward force provided by contact with the table, and there is no movement. Likewise, when a person holds an object like a barbell, the person must provide an upward force equal to the barbell's weight to keep it from falling. Mathematics shows how Sir Isaac Newton's second law of motion can explain a very complex set of observations. Scientists and mathematicians also study other conceptualizations of gravity, such as energy extraction from gravitational fields, quantum gravity, topological gravity, and supersymmetric gravity.

Properties of Gravity

Gravitational force is peculiar in that it does not depend on motion (unlike, for example, muscle forces or aerodynamic forces). The force of gravity is the same whether the object sits on a table or is allowed to fall. For an object in free fall, Newton's second law dictates: downward acceleration = net downward force ÷ mass, and if aerodynamic forces are small enough to be neglected, net downward force is equal to weight, so that downward acceleration = weight ÷ mass.

Another peculiarity of gravitational force is that it is directly proportional to mass. Therefore (weight ÷ mass) is the same for all objects; it is approximately 9.81 m/s² near the surface of Earth, called "acceleration due to gravity," generally denoted by g.

Any object accelerates as it falls downward. Starting from rest (speed = 0), its speed after t seconds will be $g \times t$. So,

$$\text{average speed} = \frac{0 + g \times t}{2} = \frac{g \times t}{2}.$$

Therefore, the distance traveled (d) can be calculated as d = average speed × t, which can be expressed algebraically as

$$d = \frac{g \times t}{2}(t) = \frac{g \times t^2}{2}.$$

This gravitational force provides a simple method for measuring a person's visual reaction time: have the subject hold a ruler at the top and let it hang vertically. Let the subject bring his thumb and forefinger near to but not touching a known reading on the ruler, ready to grab it when it falls. At a random time, let the ruler fall. Measure the distance d it fell before it was grabbed and compute t, the reaction time, from the above equation. For d in centimeters:

$$t \text{ in milliseconds} = 45.15\sqrt{d}.$$

When gravity is the only force, whether the object is moving up, down, or at an angle, its velocity vector changes continually but its acceleration vector remains constant (magnitude g, pointing downward). The distinction between the velocity and acceleration vectors is fundamental to dynamics. The space shuttle circling Earth has constant downward acceleration when it is not firing its rockets, though its velocity—never downward—changes direction continually. Mathematics allows one to calculate what its speed must be so that the change in direction would correspond to the known constant acceleration. This speed (about 17,500 miles per hour) then determines that the period of making a complete circle around Earth is approximately 90 minutes. Farther away from Earth, gravity is weaker, so that g is smaller. It is proportional to

$$\frac{1}{r^2}$$

where r is the distance from Earth's center ("inverse square law"). Taking this factor into consideration, one can determine that a circular orbit at an altitude of 22,236 miles will take 24 hours to make a complete circle. This is, indeed, where communications satellites are located, so that they would seem not to be moving as seen from the rotating Earth. Similarly, the distance to the moon's orbit can be related to its period of revolution.

The same ideas can be applied to the gravitational forces between the sun and the planets, leading to remarkably accurate descriptions of the shapes the orbits of planets can take, the change in speed as the orbit is traversed, and the relation between period of revolution and distance from the sun. All this follows from Newton's second law and a rule of how much the gravitational force weakens with distance.

Further Reading

Buchbinder, Joseph, and Sergei Kuzenko. *Ideas and Methods of Supersymmetry and Supergravity*. Oxfordshire, England: Taylor & Francis, 1998.

Gamow, George. *Gravity*. New York: Dover Publications, 2002.

Rovelli, Carlo. *Quantum Gravity*. Cambridge, England: Cambridge University Press, 2007.

Ziaul Hasan

Green Mathematics

Category: Weather, Nature, and Environment.
Fields of Study: Algebra; Data Analysis and Probability; Measurement; Number and Operations.
Summary: Modeling, analysis, and computation are used to promote environmentally conscious practices.

Green mathematics is the use of mathematical modeling, analysis, and computation to promote ecologically sound practices, such as sustainable production or reduction of pollution. Green mathematics is an increasingly popular and controversial topic with ties to other contentious social, scientific, and political issues, such as recycling laws and global warming. It is a rich area of research and development for mathematicians and scientists. For example, computer scientists Young Choon Lee and Albert Zomaya have developed and patented an Energy Conscious Scheduling (ECS) algorithm. The ECS software maps the assignment of computational tasks in high-performance computer systems as a function of the dynamic voltage scaling capability of the processors. It optimizes scheduling to decrease task completion time and energy use. Green mathematics also appeals to many mathematics educators at all levels for its apparent applicability, real-world connections, and the ability to connect to academic curriculum in other areas like history and science. In 2010, Roger Williams University's student mathematics fair was organized around the theme "Designer Math Goes Green! Mathematics and the Environment!" College programs for ecology and sustainable development rely heavily on mathematics and statistics for research and applications. On the other hand, "green math" can have negative connotations for some people, especially when it affects taxpayer dollars and restrictive changes in public policy. In some cases, this reflects an incomplete understanding regarding the basis for such calculations and the methods by which final figures are derived, often because such information is not presented to the public. In others, this may result from inappropriate extrapolation or the political "spin" attached to such calculations.

Green Measurements and Metrics

Measurements of sustainability and environmental impact apply to persons, groups, products, and events. Carbon footprint, for example, is the measure of total emission of greenhouse gases, mostly carbon dioxide, involved in an event or in the lives of people over a given time, usually a year. Energy consumption measures how much energy a process or product takes over its lifetime and accounts for sources of energy, such as atomic or fossil fuels. Ecological metrics may also include emissions of chemical pollutants, such as heavy metals, strength of potentially harmful electromagnetic fields, and intensity of light pollution. The units of measure vary by type of pollution; for example, weight per volume is used for water quality measurements, but light pollution is measured in changes in sky brightness.

Quality standards in ecological measurements include some of the same principles that apply to measurement in general, such as precision and accuracy, together constituting validity. In addition, the measurement of ecological impact requires holistic, systemic approaches, taking into account interactions among multiple variables and their relative weight for particular ecosystems. For example, different ecosystems have different resource balances. Polluting a scarce resource, such as the only water source in a desert oasis, has higher environmental impacts than polluting an abundant resource. This can be reflected in mathematics equations by applying different coefficients to different types of impacts, according to the particular situation within each ecosystem.

It is difficult to weigh different types of environmental impacts against one another. For example, producing paper from trees grown for that specific purpose takes less energy than recycling paper, but involves more water and air pollutants.

Computational Modeling in Ecology

A mathematical model is an idealized system of variables, parameters, and equations governing relationships, assumed to be close enough to a real system for the purposes of prediction or explanation. Mathematical models in ecology are typically based on observations of sets of data from real environments and involve hypothesizing about sets of data that would result if variables changed.

Models can predict developments of ecological systems if the outputs of models, taken over time, fit the corresponding changes in variables of the real ecosystem closely enough. Evaluation of a model includes its accuracy, based on a statistical metric of closeness between observed and predicted data. Nonparametric statistics is the field that deals with evaluating the accuracy of models when the data is limited and not all mathematical assumptions can be tested.

The explanatory power of a model is based on the claim that the model preserves cause-effect relationships within the ecosystem. In mathematical models, such relationships are expressed as algebraic or differential equations among variables of the model.

The possibility of general patterns (models) in ecology has to do with two global problems, or hypotheses: contingency and complexity. The contingency hypothesis says that causal relationships in any given ecosystem are so numerous that projection from one system to another is not possible. The complexity problem is that the number of variables and their weak interactions in any given ecosystem are beyond the computational power theoretically available, making systems immeasurable, their equations insoluble, and the models unable to be interpreted. That is, contingency and complexity are theoretical and philosophical challenges to the possibility and validity of ecological modeling.

Environmental Considerations by Type of Mathematics

Different areas of mathematics allow different approaches to environmental problems. Algebraic reasoning, for example, assumes functional dependencies among variables and known operations. It is most appropriate in cases where algebraic relationships among variables are stable over time and can be established with empirical measurements. For example, producing one megajoule of energy by burning coal emits 92 grams of carbon dioxide. One can compute the carbon footprint of heating a house by coal algebraically by measuring the energy consumption and multiplying it by 92 grams of coal.

Calculus is the study of rates of change in variables and limits of change. In green mathematics, calculus methods are most appropriate when algebraic relationships between variables and their changes over time are measurable. For example, rocket propulsion consumes fuel stored within the vehicle, making the vehicle lighter with time. The efficiency of rocket engines can be computed by applying integrals over time to equations connecting changes in mass and momentum resulting from the engine.

Differential equations involve the study of unknown functions by known values and their rates of change, that is, derivatives—a situation frequently encountered in ecology. Differential equations are extensively used in green mathematics to model interactions within systems, such as predator-prey dynamics, fluid dynamics in natural and human-made water and gas systems, radioactive decay, or economic growth.

Statistical methods deal with the organization and interpretation of data that include random elements. Descriptive statistics summarizes patterns in data collected from some group of objects or events, called "population." It may include data calculations such as mean or frequency. Descriptive statistics is useful for comparing systems that include randomness, such as per capita consumption of energy in different countries or recycling behaviors in neighborhoods of a city.

Inferential statistics predicts patterns in the whole population based on data observed in a sample of the population. It is extensively used in biology, ecology, and economics because collecting data about every element in the population is rarely possible. One of the most powerful methods of inferential statistics is the analysis of correlations within data. For example, the levels of air pollution in cities correlate with the incidence of asthma among the population. Notably, even strong correlations between two variables do not necessarily mean particular cause-effect relationships. The first variable may depend on the second, or the second on the third, or both may depend on another factor. For example, in children younger than 6, problem-solving abilities strongly correlate with foot size. The reason is that both foot size and problem-solving abilities increase with age.

Data visualization is an interdisciplinary area spanning descriptive statistics; grid and graph use from algebra and calculus; specific representation methods from more narrow areas of mathematics, such as tree diagrams from combinatorics; psychology of perception and learning; and design. Visual literacy combines the ability to understand and critically analyze visualizations produced by others and to create quality visualizations for the purposes of analyzing and sharing messages. Because green mathematics frequently deals with controversial issues, individuals and groups promoting different agendas use and often abuse data visualization to make their point. Visual literacy is one of the "twenty-first-century skills" whose importance is growing with heavier use of mathematics in ecology and growing emphasis on ecological approaches in all areas of life.

Green Economics and Sustainability

Mathematics is used to describe, plan, model, and predict green economy, which is economy based on ecological and social sustainability. Sustainability is a system's capacity to endure over time, measured by a variety of indices and metrics. For example, the biodiversity index measures the number of plant and animal species in an ecosystem. Using an old-growth forest for lumber and replanting trees may produce the same amount of biomass, but such "farmed" forest typically has a much-lower biodiversity index. Air quality indices assign point values to combinations of air pollutants, such as dust, ground-level ozone, and sulfur dioxide. Higher values of an air quality index correlate with higher incidents of asthma and other adverse health effects. Factories and other entities and events can be evaluated by their effects on an air quality index.

Carrying capacity of an environment, with respect to a species, is the number of individuals the environment can sustain. In differential equations, carrying capacity is the stable state of the system: populations over carrying capacities decrease over time, and populations under carrying capacities grow. Carrying capacity for humans changes depending on their practices. For example, hunter-gatherer tribes need larger areas for sustenance than groups that practice agriculture. The classic mathematical models of carrying capacity were developed for animal populations in relatively small and closed ecosystems. Because people actively change their environments, travel, and exchange resources globally, such models need significant modifications for applications to humans. Current mathematical models are based on evaluating population growth and resource use over time. For example, mining for groundwater can dramatically increase agricultural outputs and thus support population growth until the water runs out, at which time famine can lead to a population collapse.

Further Reading

Fusaro, B. A., and P. C. Kenschaft. *Environmental Mathematics in the Classroom.* Washington, DC: Mathematical Association of America, 2003.

Hanebuth, Eddie. *A Geospatial Industry Series in Science, Technology, Engineering, & Mathematics: Green & Sustainability Focus.* Ridgeland, MS: Digital Quest, 2010.

Pfaff, Tom. "Mathematics and Sustainability." http://www.ithaca.edu/tpfaff/sustainability.htm.

Maria Droujkova

Growth Charts

Category: Medicine and Health.
Fields of Study: Algebra; Measurement.
Summary: Children's development both mentally and physically is modeled using data-based norms, some of which are indicated by growth charts. .

When a parent brings his or her child to a physician for a checkup, a number of measurements are taken to help the physician assess the health and development of the child. For children up to 36 months of age, three typical measurements include height, weight, and head circumference. The healthcare professional will use these measurements to decide whether the child is on track developmentally. These measurements are expected to vary depending on the gender and age of the child. Considering weight, for instance, younger children tend to weigh less than older children and girls tend to weigh less than boys. However, there is even considerable variability in these measurements for children within the same gender and age group. There are individual differences from child to child resulting from genetic and environmental factors, including diet and physical activity habits.

Percentiles

To make a judgment about whether the child's development is on track, the relevant question to pose is where the child's measurements fit in relation to other children of the same age and gender in the population. Percentiles are typically used to facilitate this comparison and growth charts summarize these quantities in graphs. If a young boy's weight is at the 75th percentile, this means that of the boys the same age in the population, about 75% of them weigh less and about 25% of them weigh more than this boy. If parents are told that one of their child's measurements is at the 99th percentile, should they be concerned? Very high or very low percentiles may be a sign of something abnormal. For example, a child's weight at the 4th percentile may be a sign of malnutrition. Extreme measurements indicate to the healthcare professional that further follow-up may be necessary. Generally speaking, measurements under the 5th percentile or over 95th percentile or growth patterns that shift considerably in terms of their percentiles over time require further assessment.

Growth Charts

Growth charts are graphical summaries of mathematical functions that are developed based on extensive body measurement data collected on large groups of children from the population of interest. They provide benchmarks for comparison and are widely used by the health community to monitor and track the growth and development of children. According to the Centers for Disease Control and Prevention (CDC), growth charts have been used in the United States since 1977. Prior to 1977, there were child development references in use, but they did not adequately represent the population. As of 2011, the charts used in the United States are the 2000 CDC Growth Charts. The infant (0–36 months of age) charts include smoothed percentile curves of weight by age, length by age, head circumference by age, and weight by length for boys and girls. The children and adolescent (2–20 years of age) charts include weight by stature, weight by age, stature by age, and body mass index by age for each sex.

In order to find a percentile based on these charts, one needs to be able to plot a point on the graph. For instance, consider the weight-by-age infant chart for boys. Suppose the boy is 18 months and his weight is 25 pounds. Find 18 months along the horizontal axis and 25 pounds along the vertical axis of the graph. Mark this point. Based on the chart, this point falls between the 25th and 50th percentile curves. As demonstrated by this example, not all percentile curves are summarized in the charts. If a measurement falls somewhere between the 3rd, 5th, 10th, 25th, 50th, 75th, 90th, 95th, or 97th percentile curve, the professional reading the chart will need to interpolate between curves to approximate the percentile value.

The percentile curves summarized in the 2000 CDC Growth Charts were developed by the United States National Center for Health Statistics (NCHS) based on the results from a number of large health surveys conducted based on representative groups from the U.S. population. Data analysis was used to estimate percentiles for the various growth measurements and statistical modeling was used to smooth the estimates into the percentile curves to facilitate comparisons.

Further Reading

Kuczmarski, R. J., C. L. Ogden, and S. S. Guo, et al. "2000 CDC Growth Charts for the United States: Methods and Development." *National Center for Health Statistics. Vital and Health Statistics* 11, no. 246 (2002).

World Health Organization (WHO). "The WHO Child Growth Standards." http://www.who.int/childgrowth/standards/en.

Bethany White

Hurricanes and Tornadoes

Category: Weather, Nature, and Environment.
Fields of Study: Data Analysis and Probability; Geometry; Measurement; Problem Solving.
Summary: Mathematical analysis and modeling have been used to attempt to predict and simulate hurricanes and tornadoes.

Hurricanes and tornadoes are both potentially catastrophic types of storms that cause billions of dollars in damage and claim many lives each year. Predicting major weather events of these types is difficult, though mathematical modeling and computer power have allowed mathematicians and scientists to make advances in storm science. The term "cyclone" is often erroneously applied to tornadoes; it properly refers to the class of storms originating over water that includes hurricanes, typhoons, and tropical cyclones. Tornadoes and cyclones are characterized by revolving forms and high winds, but tornadoes are typically smaller, faster spawning, shorter lived, and their damage is usually more focused. Mathematical analysis and modeling of storms draws from many fields.

For example, vector calculus plays a substantial role in analyzing and modeling these storms, since both pressure and humidity can be represented as scalar fields and wind as a vector field. Theories and equations from physics for conservation of mass and energy, along with angular momentum and shear, are also quite important. Historically, challenges in storm description, prediction, modeling, and simulation have often been related to data collection and computing power. One of the earliest systematic data collection and prediction efforts was conducted in the 1880s by John Finley of the U.S. Army Signal Corps, but for a variety of sociopolitical reasons, federal research lagged until about World War II. The emergence of Doppler radar advanced storm science, as did computers in the 1970s that were capable of generating three-dimensional models. However, even in the twenty-first century, no one can perfectly predict the emergence, path, strength, or damage of a hurricane or tornado. Even with multiple stations and satellites, data are still sometimes sparse or difficult to integrate across sources, and this type of research raises theoretical questions about the limits of predictability. At the same time, early warning systems that give even a few hours of notice regarding approaching storms are widely considered to be beneficial, and mathematicians continue to contribute to this area. Actuaries are also involved in calculating the costs of these storms, in terms of both money and lives.

A tornado is a rotating column of air that is in contact with both the ground and a cloud. Tornadoes are generally spawned by thunderstorms. The United States has the highest incidence of tornadoes of any country in the world, in part because of the confluence of cold air from Canada, warm, moist air from the Gulf of Mexico, and dry air from the Southwest. A related phenomenon is water spouts, which are essentially tornadoes that form over water, especially in tropical areas. A hurricane is a powerful, spiraling storm that begins over a warm sea, near the equator. "Hurricane" is, in fact, just one name for the kind of storm scientists refer to as a "strong tropical cyclone." Depending on where they occur, hurricanes are given a different label. If they begin over the Atlantic Basin (Atlantic Ocean north of the equator, the Caribbean Sea, the Gulf of Mexico) or the Northeast Pacific Ocean, they are called "hurricanes."

When the same kind of storm occurs in the western North Pacific Ocean, it is called a "typhoon." In the southwest Pacific Ocean and the Indian Ocean, the storms are referred to as "cyclones." No matter what it is called when a hurricane, typhoon, or cyclone hits land, it can do great damage through fierce winds, torrential rains, inland flooding, and huge waves crashing ashore. A powerful hurricane can kill more people and destroy more property than any other natural disaster. Hurricanes and other cyclones form in the tropics during summer and fall.

Predicting Major Storms

A few very important characteristics of hurricane are as follows:

- Hurricanes form under weak, high-altitude winds
- Hurricanes have no fronts
- Hurricanes main energy source is the latent heat of condensation
- The center of a storm is warmer than the surrounding air
- Hurricane winds weaken with height
- Strongest winds are near the Earth's surface
- Hurricanes weaken rapidly over land

As global weather patterns become more erratic as evidenced in the early twenty-first century, it has become difficult to accurately forecast hurricanes. However, mathematics allows forecasters a thorough insight into the mechanisms of weather features, including large-amplitude water waves and sustained winds cloud structure. Moreover, statistical models built from historical data perform with greater precision. Also, scientists use high-quality time series data along with less precise time series data using a Bayesian approach, which does not require data to have uniform precision. This way, scientists have been able to forecast U.S. hurricanes six months in advance.

Wind engineer Herbert Saffir and meteorologist Robert Simpson introduced the very popular Saffir–Simpson wind scale, which is a 1–5 categorization based on the hurricane's intensity at the indicated time. This scale is an excellent tool for alerting the public about the possible impacts of various-intensity hurricanes. However, the scale does not address the potential for other hurricane-related impacts, such as storm surges, rainfall-induced floods, and tornadoes.

The estimation of hurricane-generated waves and surges in coastal waters is of critical importance to the timely evacuation of coastal residents and the assessment of damage to coastal property in the event that a storm makes landfall. Tornado wind speed or intensity is rated using the Fujita scale, named for Tetsuya

A weather satellite image of Hurricane Katrina in the Gulf of Mexico. Starting as a slight pressure difference, hurricanes grow into large spiraling storm systems of low pressure, complete with high winds and driving rain. (National Aeronautics and Space Administration)

Theodore Fujita. It is based on the subjective assessment of the damage caused to human and vegetation structures by the tornado. Its original development was linked to the Beaufort wind force scale, named for Francis Beaufort. Ratings range from a minimum of "F0" to a maximum of "F6." It is also sometimes called the Fujita–Pearson scale to recognize contributions of Allen Pearson, who was director of the National Severe Storms Forecast Center at the time. The scale has since been revised by data gathered from structural engineers and others that suggested that the original wind speeds were too high for categories F3 and above.

To provide accurate estimates for wave height, scientists use Wave Model (WAM). WAM is built around the solution to the action balance equation in terms of an action density function. With the aid of FORTRAN and other programming languages today, WAM is an extremely efficient model.

Hurricane size (extent of hurricane-force winds), local bathymetry (depth of near-shore waters), topography, the hurricane's forward speed, and its angle to the coast are all factors that affect the surge that is produced. Mathematicians and scientists are working hard to develop a reliable technique for prediction of storm surges. The capability for prediction of hurricane surges is based primarily on the use of analytic and mathematical models, which estimate the interactions between winds and ocean, also taking into account numerous other factors. One of the models used for storm-surge modeling is known as the Advanced Circulation Model (ADCIRC). This is a finite-element circulation model based on the two-dimensional, depth-integrated shallow-water equations representing the conservation laws for mass and momentum. The momentum equations are combined with the continuity equation and result in the generalized wave continuity equation.

ADCIRC is implemented in spherical coordinates for this application. As expected, many parameters can be set to optimize running the model for specific applications and locations.

Further Reading

Adam, John. *Mathematics in Nature: Modeling Patterns in the Natural World.* Princeton, NJ: Princeton University Press, 2003.

Elsner, J. B., et al. "Bayesian Analysis of U.S. Hurricane Climate." *Journal of Climate* 14, no 23 (2001).

Kumer Pial Das

Infectious Diseases, Tracking

Category: Medicine and Health.
Fields of Study: Communication; Data Analysis and Probability.
Summary: Physicians and mathematicians have long worked together to develop and use models that track the spread of infectious diseases in order to develop appropriate countermeasures and responses to halt the disease spread.

The health of societies relies on quickly and correctly tracking and predicting the growth and spread of disease in populations. Epidemiology is a mathematically rich area. Exposure and infection are both probabilistic processes, and tracking infectious diseases is a dynamic application of mathematics. The World Health Organization (WHO) and other organizations concerned with public health use mathematical models in their decision-making, such as when WHO analyzed the risks and benefits of travel restrictions during the early twenty-first-century H1N1 (swine flu) epidemic. Epidemiology has a long history with important societal connections. Some trace one early use of mathematical modeling for disease to eighteenth-century mathematician Daniel Bernoulli. He presented an analysis of smallpox morbidity and mortality to demonstrate the efficacy of vaccination.

Nineteenth-century physician William Farr is often called the "father of epidemiology" and was responsible for the collection of official medical statistics in England and Wales. His most important contribution was to set up a system for routinely recording causes of death. Physician John Snow is frequently cited as using graphical methods to propose a mechanism of transmission and the source of a cholera epidemic in nineteenth century London. Epidemiologists using mathematical and statistical models have been influential in research, treatment, and some methods of prevention for potentially devastating diseases, like tuberculosis, smallpox, typhus, and malaria.

Infectious diseases are a leading cause of death for humans. In order to understand the dynamics of tracking infectious disease at the population level, it is important to understand the responsible mechanisms at the individual level. Infectious disease is caused by a pathogenic agent (for example, a virus, bacterium, or parasite) transmitted through one of many methods, such as air or body fluids. One method scientists have developed for investigating why outbreaks of disease take place and how to contain or end them is to design a system of surveillance and data collection from individual cases, which can then be used to model the infection's trajectory through a population. Other times, they may use data from past similar situations to extrapolate possible solutions.

Surveillance of Infectious Disease

Central public health institutions have created computer systems to monitor emerging outbreaks of illnesses. Traditional notification has relied on disease reporting by laboratories and hospitals. However, the first indications of an outbreak usually occur before a formal diagnosis. People respond to illness with a variety of behaviors to illness that can often be tracked; for example, the number of visits to emergency rooms, or purchases of over-the-counter drugs. Other people's behaviors are more difficult to track, such as those people who continue their daily routines even though they feel sick. Systems of surveillance may compile data from many sources to look for unusual patterns or significant increases in activities like emergency room visits. Another approach, based on Internet search queries, collects disease-related searches. The searches are linked to geographic mapping tools and are used to identify clusters of symptoms. Further analysis and modeling using mathematical and statistical methods are needed to estimate the potential impact of a disease outbreak.

This microfluidic labchip was used in a CDC bioanalyzer to evaluate Mycobacterium cosmeticum *strains. There are some 115 species of* Mycobacterium, *causing infectious diseases like tuberculosis and leprosy.* (Center for Disease Control)

Modeling Infectious Disease

Quantitative analysis describes probable disease trajectories for predicting impact over time. The parameters may include the variables of time, geographic location, population density, contact rate, and saturation, as well as the personal characteristics of those who contract the disease. For example, eighteenth-century mathematician Daniel Bernoulli created mathematical models for smallpox to support the use of inoculations. At the turn of the twentieth century, British physician Ronald Ross began to develop mathematical models to help him understand malaria's trajectory, rate of progression, and probability of infection. He received the Nobel Prize in Physiology or Medicine in 1902, indicating the importance of his mathematical contributions to epidemic theory. Another early twentieth-century model is the Reed–Frost epidemic model, which was developed by scientists Lowell Reed and Wade Hampton Frost. It models disease transmission via person-to-person contact in a group and includes concepts like a fixed probability of any person coming into contact with any other individual in the group.

Quantitative research continued throughout the twentieth century and continues to be active in the twenty-first century. There are many large agencies that use epidemiological models, such as WHO and the U.S. Centers for Disease Control and Prevention (CDC). As medicine and technology advance, new variables become important in models; for example, global air travel, which brings previously isolated populations into greater contact with one another, along with new vaccinations and vaccination policies. Differential use of longtime practices, like quarantining sick and potentially exposed individuals, may also be a factor.

Other models incorporate seasonal information, such as varying contact rates, which can be affected by societal structures, such as school schedules. In the latter twentieth century, computer networking and the subsequent spread of computer viruses have led mathematicians and others to extend epidemiological models to research and model the spread of computers worms and viruses using mathematical techniques, such as directed graphs and simulation. In such an active field of research, new technologies and methods for quickening the pace of identifying patterns of disease are expected to be developed.

Further Reading

Diekmann, O., and J. A. P. Heesterbeek. *Mathematical Epidemiology of Infectious Diseases: Model Building, Analysis and Interpretation*. Hoboken, NJ: Wiley, 2000.

Hethcote, H. W. "The Mathematics of Infectious Diseases." *Society for Industrial and Applied Mathematics* 42 (2000).

Keeling, Matt, and Pejman Rohani. *Modeling Infectious Diseases in Humans and Animals*. Princeton, NJ: Princeton University Press, 2007.

Douglas Rugh

Intelligence Quotients

Category: Medicine and Health.
Fields of Study: Algebra; Geometry; Number and Operations; Problem Solving.
Summary: Intelligence tests are created and analyzed using mathematics.

The term "intelligence" is broadly synonymous with the term "cognitive ability." Intelligence tests are tests designed to measure cognitive abilities. According to Ian Deary and David Batty, cognitive abilities are mental abilities "that are not principally sensory, emotional or conative (related to the will)." Standardized intelligence tests produce a score called the "Intelligence Quotient" (IQ). IQ tests are usually copyrighted, and to prevent people from practicing for them, they must be administered in supervised conditions. Many tests that claim to measure IQ have appeared on the Internet but may not have been validated by professional psychologists. Intelligence, or cognitive ability, has been defined in different ways but broadly refers to people's ability to process complexity "on the spot."

Since psychologists such as Alfred Binet (originator of the test that later evolved into the Stanford–Binet) and David Wechsler (creator of the Wechsler Adult Intelligence Scale and Wechsler Intelligence Scale for Children) began measuring cognitive abilities over 100 years ago, nearly all measures of cognition have been shown to correlate. This fact is interpreted as evidence for a general factor, called g, representing general intelligence. At the beginning of the twenty-first century, no test of cognitive ability has been created that does not correlate with other cognitive ability tests. In practice, this means that people who are good at processing complexity in one area tend to be good at processing complexity in another. A person's IQ score is a numerical representation of their level of g.

Most IQ tests are designed to have a mean of 100 and scores are normally distributed. However, the standard deviation varies across different tests. The interpretation of the standard deviation is that it represented the average distance from the mean, in either direction. To understand and interpret a person's IQ score, it is necessary to know the standard deviation of the test they took. Common standard deviations are 15 or 16, and the range of IQ scores is generally between about 55 and 145 for a test with a standard deviation of 15. Further, about two-thirds of individuals will have scores within one standard deviation of the mean and about 95 percent will have scores within two standard deviations of the mean. For this reason, IQ scores are sometimes evaluated using percentile scores, which divide the normal distribution into 100 parts so that 1% of the scores are in each part. For example, admission to the high-IQ society Mensa requires a person to score in the 98th percentile or higher on several different validated IQ tests. This requirement means about one in 50 people would be eligible to join.

Percentile IQ scores can be useful, but they can be misinterpreted since the distance between each percentile is not equal. In contrast, standard deviations are the same distance apart, sometimes making it more sensible to compare individuals in terms of average distance from the mean. Also, IQ tests are imperfect measures of intelligence because they generally do not produce the exact same score for the same person, even if the test is taken more than once. This inaccuracy is quantified by the standard error of measurement and represents how much variability an individual person's scores would have if they took the test many times. For example, if a person scored 100 on an IQ test that had a standard error of 2, the person's true IQ score would often be interpreted as being somewhere between 96 and 104. Some researchers and others have suggested that the average of three IQ tests provides a better indication of a person's true IQ score than a single test.

There are three features of general intelligence that are important because they negate arguments that IQ scores have no meaning: their stability, their heritability, and their correlation with external phenomena. First, IQ scores are remarkably stable across the life course from childhood to old age. Data to demonstrate this are exceptionally rare, but one exception can be found in Scotland. During one day in 1932, every 11-year-old in the country took an IQ test. They were retested 66 years later, and the scores were found to correlate highly with childhood IQ score (0.76), providing evidence of stability of IQ scores over time. Second, IQ scores are highly heritable. The heritability of individual differences has been estimated as between 30% and 80%, illustrating that genetics contributes strongly to IQ scores. However, no single gene or set of genes has been identified. This suggests that the genetic contribution to intelligence is multifactorial, as with other observable characteristics (phenotypes), such as height. There are no sex differences in IQ, although the distribution of males' scores is slightly wider at both ends of the distribution. Third, IQ scores correlate with variables that can be considered external, or outside the IQ test itself. IQ correlates with indicators of socioeconomic status (SES)—a indication of factors like educational attainment, income, and occupational social class—and with many biological variables, including brain size, height, sperm quality, and mortality. The causes of these correlations are disputed.

Content of IQ Tests

The content of IQ tests differs, depending on the specific cognitive abilities they are intended to measure. Some tests have been criticized as being culturally biased because they ask questions that require culturally specific knowledge. Tests that do not evaluate "general knowledge" are considered more "culture fair." For example, Raven's Matrices is a test that contains no written information, requiring abstract reasoning skills. This test contains no culturally specific information, so that it is not possible to learn how to take the test. Similarly, tests of reaction time are considered indicators of g, because they reflect speed of information processing. These do not assess culturally specific information or knowledge. Clifford Pickover imagined how aliens might test human intelligence and designed related mathematics and logic puzzles. Other intelligence researchers argue that knowledge is a reliable indicator of g and should therefore be included in IQ tests. IQ tests also differ in the extent to which it is necessary to complete every question. Traditional IQ tests are designed using classical test theory. In these tests, the IQ score is more reliable for people with an average level of IQ. Since people with high IQs find many questions easy and people with low IQs find many questions difficult, fewer relevant questions are answered by people at either end of the IQ distribution. More recently, computerized adaptive tests have been developed and informed by item response theory, which addresses these problems. These tests can alter the difficulty of test items, so that people with high IQs receive a larger number of difficult items. Reliability is improved and testing length can be reduced because respondents do not have to answer every question.

Age and Intelligence

Although IQ scores are relatively stable, cognitive decline typically occurs with increasing age. This fact is important because cognitive decline may indicate mild cognitive impairment and risk of dementia. When considered over time, specific kinds of cognitive abilities appear to deteriorate at different rates. Fluid intelligence, referring to processing speed (particularly of new information), declines from age 26 onward. In contrast, crystallized intelligence, referring to speed of recall of existing knowledge (for example, vocabulary and general knowledge) is relatively stable.

For this reason, standardized tests of word recognition, such as the National Adult Reading Test (NART), are useful at estimating premorbid IQ in patients suspected of having dementia. A discrepancy between IQ as estimated by the NART and IQ estimated from another test could indicate that cognitive decline has occurred. Cognitive decline can result in mild cognitive impairment and dementia or Alzheimer's disease, which have high mortality, morbidity, and treatment and care costs.

Research into the prevention of cognitive decline is ongoing, but several risk factors have emerged consistently, such as cigarette smoking and physical inactivity. Consumption of fish oils, either from oily fish or fish oil supplements, may help prevent cognitive decline. Prior IQ is a strong protective factor, such that a higher initial IQ appears to protect against cognitive decline in later life. Claims that IQ can be changed are controversial. Although brain plasticity is known to be greater than once thought—and there is evidence that children exposed to cognitive stimulation enjoy increases in IQ—it is not clear how stable these gains are. Furthermore, attempts to increase IQ in adults have not been successful.

Lower IQ scores are associated with earlier mortality and higher morbidity. This association provides further evidence for the validity of IQ tests. It is noteworthy that the relationship between IQ and mortality often remains after adjusting for indicators of socioeconomic status (SES), such as income, educational attainment, and occupational social class. Given that IQ is largely stable after childhood, this relationship is unlikely to be explained by societal factors. Evidence suggests that IQ contributes strongly to health literacy, which, according to the World Health Organization, refers to "the cognitive and social skills which determine the motivation and ability of individuals to gain access to, understand, and use information in ways which promote and maintain good health." People with inadequate health literacy skills tend to have unhealthier lifestyles, adhere less well to medical regimens, and do not understand written health information or the need for regular screening for diseases.

Access to healthcare does not solely explain the IQ-health relationship because it can also be found in countries that have free healthcare, such as the United Kingdom, which has the National Health Service (NHS). Managing chronic diseases, such as diabetes, involves repeating many complex tasks, such as monitoring blood sugar and planning activities around meals. Without supervision and support, the risk of making dangerous mistakes could accumulate over time. Many areas of life involve repeating a set of unpredictable, complex tasks, which can damage health in the long term. The field of cognitive epidemiology studies why IQ is linked to worse health outcomes and the role that literacy plays in this relationship.

Further Reading

Deary, Ian. *Intelligence: A Very Short Introduction*. Oxford, England: Oxford University Press, 2001.

Deary, Ian, Lars Penke, and Wendy Johnson. "The Neuroscience of Human Intelligence Differences." *Nature Reviews Neuroscience* 11, no. 3 (2010).

Gottfredson, Linda. "Intelligence: Is It the Epidemiologists' Elusive 'Fundamental Cause' of Social Class Inequalities in Health?" *Journal of Personality and Social Psychology* 86, no. 1 (2004).

Gottfredson, Linda, and Ian Deary. "Intelligence Predicts Health and Longevity, But Why?" *Current Directions in Psychological Science* 13, no. 1 (2004).

Gareth Hagger-Johnson

Interplanetary Travel

Category: Space, Time, and Distance.
Fields of Study: Geometry; Measurement; Problem Solving.
Summary: Space exploration requires mathematics to plan trajectories and to navigate in space, as well as to measure and to analyze massive amounts of data.

Interplanetary travel can be defined as any spaceflight—manned or remotely guided—to the various bodies of the solar system, including planets, their satellites, and asteroids. Such space exploration required new mathematics to plan trajectories and navigate in space, as well as to measure and to analyze massive amounts of data. These flights have had a great societal impact and have radically changed human attitudes toward the outer space surrounding the Earth.

History

A scientific possibility of interplanetary travel was discussed for centuries after Isaac Newton wrote *Principia* in 1687, in which he unified terrestrial and celestial dynamics by discovering the force of gravity as an important source of motion, including the movement of celestial bodies. Step by step, an important new mathematical branch of astronomy emerged and received the title "celestial mechanics." In its formative days, celestial mechanics played an outstanding role in the progress of mathematics, demanding and inspiring novel and efficient mathematical tools. Among the pioneers of celestial mechanics were prominent mathematicians such as Leonhard Euler (1707–1783), Alexis-Claude Clairaut (1717–1765), and Joseph-Louis Lagrange (1736–1813). Today, the branch of celestial mechanics dedicated to spaceflight is usually termed astrodynamics.

For many years following Newton's discovery, the topic of interplanetary travels mainly remained the subject of science fiction writers. In the nineteenth century, among the most influential science fiction writers were Jules Verne (1828–1905) with his books *From the Earth to the Moon* and *All Around the Moon* and H. G. Wells (1866–1946) with his book *War of the Worlds*. Verne's work contained a great deal of mathematics discussion, much of which was reasonably accurate based on the knowledge of the time.

To put interplanetary travel into practice, it was necessary to realize some significant preconditions, including designing spacecraft with the capacity for maneuvering, designing technologies for boosters to reach escape velocity, developing a theoretical base for space navigation, and creating systems for long-distance radio communications. These technological developments were not made until the beginning of the space era in 1957.

Mathematical Development

From a mathematical viewpoint, the most interesting part of interplanetary travel is space navigation. An appropriate example of a solution with respect to navigational problems is the Hohmann transfer orbit. In 1925, Walter Hohmann calculated that the lowest-energy route between any two celestial bodies is an ellipse that forms a tangent to the starting and destination orbits of these bodies. Such a transfer orbit between the Earth and Mars is graphed in the following illustration. A spacecraft traveling from Earth to Mars along the Hohmann trajectory will arrive near Mars's orbit in approximately 18 months. Just a small application of thrust is all that is needed to put a space probe into a circular orbit around Mars. The Hohmann transfer applies to any two orbits, not just those with planets involved (see Figure 1). In the figure, Hohmann Transfer Orbit (light gray oblong ring), Earth's orbit is represented by

Figure 1. Hohmann Transfer Orbit.

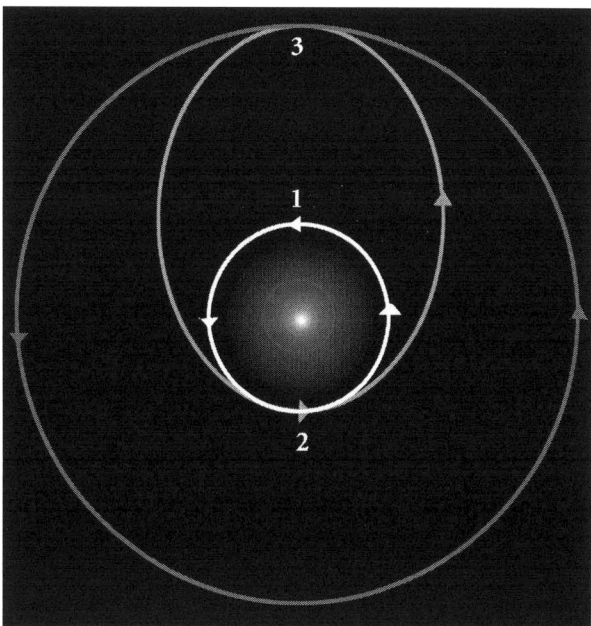

the white circle, and Mars' orbit is represented by the darker gray circle. A spaceship leaves from point 2 in Earth's orbit and arrives at point 3 in Mars's.

Another example of navigational technique is routinely called the "gravitational slingshot." It utilizes the gravitational influence of planets and their moons to change the speed and direction of a space probe without the application of an engine. In this case, a spacecraft is sent to a distant planet on a path that is much faster than the Hohmann transfer. This would typically mean that it would arrive at the planet's orbit and continue past it. However, if there is a planetary mass between the departure point and the target, it can be used to bend the path toward the target, and in many cases the overall travel time is greatly shortened. Prime examples of the gravitational slingshot are the flights of the two spacecraft of the American Voyager program, which used slingshot effects to redirect trajectories several times in the outer solar system. Astrodynamics considers many other interesting approaches. Several technologies have been proposed that both save fuel and provide significantly faster travel than Hohmann transfers; most are still theoretical.

Because of astrodynamics limitations, travel to other solar systems bodies is practical only within certain time windows. Outside of such windows, these bodies are essentially inaccessible from Earth using current technology. Mathematicians helped design the Interplanetary Superhighway, a network of low-energy trajectories, in order to find efficient routes through space; these mathematical foundations originated with French mathematician Henri Poincaré.

Achievements and Obstacles

The modern accomplishments in interplanetary travels are extraordinary. Remotely guided space robots have flown past all of the planets of the solar system from Mercury to Neptune, and the National Aeronautics and Space Administration's (NASA's) spacecraft New Horizons is scheduled to fly past the dwarf planet Pluto in 2015. The five most distant spacecraft (including the American ships *Pioneer-10*, *Pioneer-11*, *Voyager-1*, and *Voyager-2*) were scheduled to leave the solar system at the beginning of the twenty-first century. Artificial satellites have orbited Venus, Mars, Jupiter, and Saturn. Spacecraft have landed on the Moon, Venus, Mars, Saturn's moon Titan, and asteroid 433 Eros. The first probes to comets (European *Giotto*, Russian *Vegas*, American *Stardust*) were fly-by missions. In 2005, the *Deep Impact* probe hit the comet 9P/Tempel to study the composition of its interior.

Great achievements took place in manned interplanetary travels once mathematicians, scientists, and engineers understood the mathematical principles required to launch spacecraft outside Earth's atmosphere and to maneuver in the microgravity environment of space. NASA also recruited astronauts with strong academic credentials in science and mathematics. America's Mercury and Gemini programs put humans into space and Earth orbit and taught them how to change trajectory in space to move to a new orbital altitude or to dock with other spacecraft, while the Apollo program took them to the moon. After missions in which men orbited the moon and returned, Apollo 11 landed astronauts Neil Armstrong and Edwin "Buzz" Aldrin on the moon in 1969. There were six successful manned American expeditions to the moon from 1969 to 1972.

Further development of interplanetary travel has many obstacles that will require a great deal of mathematical analysis to model, simulate, and solve. For example, astronauts must be protected from extreme radiation exposure in the Van Allen belt, a torus-shaped region of space surrounding the Earth and other planets named after geophysicist James Van Allen of Iowa.

The larger outer radiation belt is about four Earth radii (RE) above the surface of the Earth and the inner is about 1.6 RE, with a gap at roughly 2.2 RE. Apollo astronauts were briefly exposed to this radiation on trips to the moon. Conspiracy theorists who disputed the notion that humans landed on the moon cited the Van Allen belt as evidence that the astronauts would have died from radiation, but simple calculations and the data collected by radiation sensors worn by astronauts (similar to those worn by scientists and hospital workers who may be exposed to radiation) demonstrated that the speed and design of the Apollo capsules protected astronauts during these relatively short trips.

If the Earth was the main focus of many sciences (geodesy, geology, geophysics, geochemistry, and oceanography) for millennia, interplanetary travel created a new important branch of research—comparative planetology—which is essential for understanding the history of Earth and its evolution.

Among many other difficult problems of interplanetary travel is developing adequate human life support. A breathable atmosphere must be maintained, with adequate amounts of oxygen, nitrogen, controlled levels of carbon dioxide, trace gases, and water vapor. It is also necessary to solve the problem of food supply.

At some point in time, all of these problems may be overcome. Incentives for future expansion of interplanetary flights include the possibility of colonizing other portions of the solar system and utilizing resources.

Further Reading

Battin, Richard. *An Introduction to the Mathematics and Methods of Astrodynamics*. New York: American Institute of Aeronautics and Astronautics, 1999.

Benson, Michael. *Beyond: Visions of the Interplanetary Probes*. New York: Harry N. Abrams, 2003.

Kemble, Stephen. *Interplanetary Mission Analysis and Design*. Berlin: Springer, 2006.

Launius, Roger D. *Frontiers of Space Exploration*. Westport, CT: Greenwood Press, 2004.

Launius, Roger D., and Howard E. McCurdy. *Robots in Space: Technology, Evolution, and Interplanetary Travel*. Baltimore, MD: Johns Hopkins University Press, 2008.

Zimmerman, Robert. *Leaving Earth: Space Stations, Rival Superpowers, and the Quest for Interplanetary Travel*. Washington, DC: J. Henry Press, 2003.

Alexander A. Gurshtein

Joints

Category: Medicine and Health.
Fields of Study: Algebra, Geometry.
Summary: Joints allow bones to move—a movement that is modeled and analyzed using mathematics.

A joint (where bones join) generally allows motion of those bones relative to each other. The motion, typically, is a rotation about the joint. Such rotations underlie almost all the movements humans perform in everyday life. Mathematics plays a crucial role in understanding the causes and consequences of the joint rotations, singly or in combination, and also in estimating the forces to which the joints are subjected.

Simple Joint Movement

Suppose, for simplicity, that rotation is confined to the elbow joint. Then the forearm would move in a plane, and the position of the hand would be represented by extrinsic (x, y) coordinates that involve trigonometric—sine and cosine—functions of the elbow angle. When many joints participate, such as the shoulder, elbow, and wrist, the description of a hand movement, like reaching for a cup, involves combinations of trigonometric functions of the joint angles. The relationship between changes in the joint angles and the resulting changes in the extrinsic coordinates is expressed in the form of a matrix (called the "Jacobian matrix"), consisting of rows and columns of trigonometric functions. The methods of matrix algebra can be used for understanding the consequences of a sequence of changes in joint angles.

The inverse problem of finding the joint angles when the extrinsic coordinates are given can have an infinite number of solutions, called "kinematic redundancy." For example, there are many ways of configuring an arm so as to get a finger to touch one's nose. Why a person chooses a certain configuration is not known, though various hypotheses have been proposed. This is a crucial issue also in robotics, where "joint" angles have to be computed in order to reach a prescribed position in space. Various mathematical methods have been utilized for picking an "optimal" solution to this problem.

Three-Dimensional Joint Movement

The importance of mathematics in understanding and describing joint function is further emphasized when considering motions in three-dimensional space

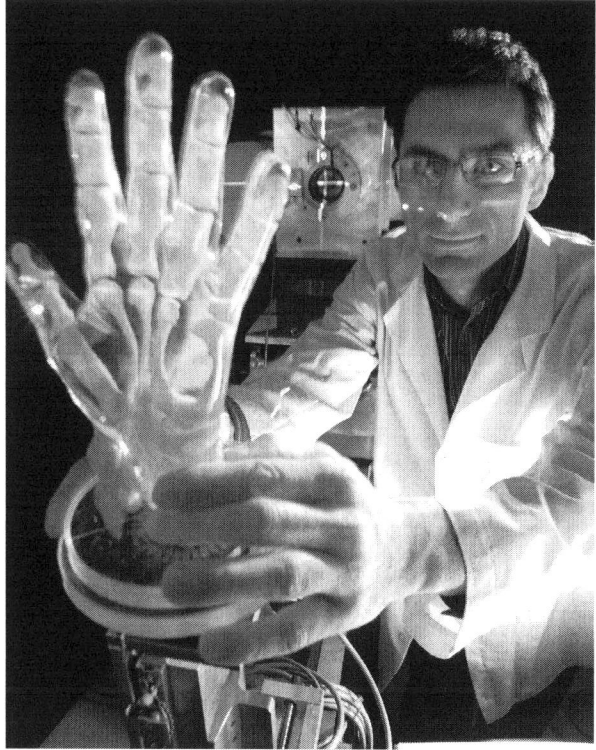

A National Institute of Biomedical Imaging and Bioengineering (NIBIB) medical imaging study. (National Institues of Health)

because certain phenomena arise that are far from intuitive. As an example, assume the shoulder to be a ball-and-socket joint and imagine the following two sequences of 90-degree rotations about the right shoulder, starting each time with the arm horizontal and stretched out to point to the right:

- Rotation about the vertical axis (bringing the arm to point to the front), followed by rotation about the left-right axis (raising the arm up, above the head)
- The opposite sequence of rotations, first about the left-right axis (twisting the arm about its long axis), followed by rotation about the vertical axis (bringing the arm pointing to the front)

The two sequences lead to different configurations. The dependence of the final outcome on the sequence of the rotations is expressed by mathematicians as the "noncommutativity" of rotations in three-dimensional space. It means that rotations can not be described simply by three numbers, unless the sequence is also specified. Certain ways of specifying the sequence have been standardized, such as a rotation being described by three "Euler angles" (yaw, pitch, and roll). There are also several other mathematical techniques, involving matrices, for dealing with rotations in 3-dimensional space, as matrices too have the property of noncommutativity ($A \times B \neq B \times A$). Another technique, which uses four rather than three numbers to represent a rotation, is the method of "quaternions." These abstract entities were proposed originally as extensions of complex numbers. Incidentally, the designers of computer visualizations, like video games, utilize quaternions for programming the rotational motions of the objects.

Forces

Motions about joints result from muscle and external forces. It is the moments of these forces that matter for rotation. In multijoint movements, a muscle moment about one joint can cause motions about several joints; specifically, even a fully relaxed joint would flop when there is motion about nearby joints. This phenomenon is described by rather complicated differential equations, which the neural control system takes into account in its planning. But the force with which the bones at a joint push against each other cannot be determined simply from the moments of forces. This force (called "joint loading") depends upon both external and muscle forces, and is typically many times greater than any external forces. The wear and tear of the joint—natural or artificial—depends upon the loading. Also, joints being nearly frictionless, slippage occurs if the load has a substantial component parallel to the surface of contact. Noninvasive techniques for estimating the joint loading force are highly computational. Given the external forces and observed motions, one determines the needed muscle torques at each joint, and then, knowing the anatomical layout of the muscles and their strengths, one estimates the distribution of forces among the muscles. With all other forces thus known or estimated, one can derive the joint loading.

Further Reading

Alexander, R. McNeill. *Principles of Animal Locomotion*. Princeton, NJ: Princeton University Press, 2006.

Burstein, Albert H., and Timothy M. Wright. *Fundamentals of Orthopaedic Biomechanics*. Baltimore, MD: Williams & Wilkins, 1994.

Hanson, Andrew. *Visualizing Quaternions*. San Francisco: Morgan Kaufmann, 2006.

Hasan, Ziaul, and James S. Thomas. "Kinematic Redundancy." *Progress in Brain Research* 123 (1999).

Ziaul Hasan

LD50/Median Lethal Dose

Category: Medicine and Health.
Fields of Study: Algebra; Data Analysis and Probability.
Summary: The median lethal dose of a compound is determined through experiment and statistical estimation.

Toxicity often needs to be compared across various chemical compounds and other noxes. The detailed and complex dose-response curve describes the relationship between the dose of a compound and its harmful effect. Frequently, a simple summary in the form of a single number is needed for practical purposes. Median lethal dose, or "LD50" is most popular in this context. It is defined as the dose at which 50% of exposed individuals die.

In order to be meaningful, such a definition implicitly assumes certain features of the dose-response relationship, namely its monotonicity, the fact that mortality increases with dosage. Although the concept is defined for a theoretical dose-response curve, its practical application is strongly related to statistical estimation of the dose-response curve model based on data obtained from an experiment with many animal or other nonhuman organisms randomly assigned to various doses.

Toxicological Testing

In toxicology and related disciplines, such as food safety and environmental risk assessment, one often needs to quantify how toxic or dangerous a substance is. A quantification of the harmful effect is needed for many practical comparisons; for instance, to compare the toxicity of different substances or to compare them with a standard. Although there are many possible aspects of how dangerous a compound is, survival of exposed individuals is frequently of interest. The survival is assessed experimentally in the "quantal response trial."

It is based on a set of animal or other nonhuman organisms, whose randomly selected groups are exposed to different doses of the tested compound. The outcomes are summarized as the percentages or proportions of those that survived in each dose group. Mortality would be just the complement of the proportion of survivors, expressed as: Mortality = (number of individuals dead after exposure) ÷ (total number of exposed individuals).

Based on common sense, one would expect that the mortality would increase with the dose of a toxic compound. Most typically this is indeed the case and the mortality obtained from an experiment with a large total number of exposed is monotonic, meaning it increases with the dose.

Dose-Response Curve

When mortality is taken as a function of dose, one can plot the so-called dose-response curve. Dose-response curve has lower asymptote at 0 since no exposure-related death can occur when no exposure is applied. Similarly, it has upper asymptote at 1, since exposure-related death will always occur with a large enough dose. The asymptotes are shown as horizontal dashed lines.

Note that often one needs to go over several orders of magnitude of doses in order to observe transition from zero effect to the full effect, so the dose-response is then plotted as the mortality versus logarithm of the dose. Since the logarithm is a one-to-one function, nothing is lost by the transformation, and the plot is more readable.

LD50

Because the dose-response curve is a rather complex quantity, many possible features might be compared across different compounds. It might be cumbersome in practice to compare curves, however. A simple summary is often all what is needed. Median lethal dose, or LD50, is the most popular characteristic. It is defined as the dose at which 50% of exposed individuals die. When a dose-response curve is available, an LD50 is constructed by drawing a horizontal line at 0.5, finding its intersection with the dose-response curve, drawing vertical line at the intersection, and reading off the value where it crosses the horizontal axis.

Statistical Estimation

In practice, one does not have the dose-response curve at hand. It needs to be estimated from experimental data by statistical means. In fact, the mortalities obtained from two experiments with the same doses would be very likely somewhat different, as a result of random errors. For example, different randomly selected experimental animals would react differently to a given dose.

Nevertheless, when the size of the experiment increases, increasing both the number of animals in every dose group and increasing the number of different dose groups, random errors would tend to decrease in line with the law of large numbers. In fact, for a very large experiment, the mortality estimates get close to the probabilities of survival. Since not all of the infinite possible doses can be explored in a real experiment, a model relating the survival probability to the dose is assumed in order to be able to interpolate between the doses actually used in the experiment. An interpolation is typically needed when calculating LD50. Parameters of the model are then estimated by various statistical means. Very often, logistic regression is used to this end.

Other Uses of LD50

While the definition of LD50 is directly related to lethality, the mathematical concepts used in LD50 testing and modeling can be applied to many other less-dramatic outcomes. In general, these models are useful when the relationship being explored involves a binary response variable, like yes/no or pass/fail, predicted by a quantitative explanatory variable, as long as the relationship is bounded and monotonically increasing in the same manner as before. For example, rather than finding the dose that induces mortality, researchers may wish to model what dose of a medicine will cause 50% of exposed individuals to show a certain, nonlethal symptom.

Further Reading

Agresti, A. *Categorical Data Analysis*. Hoboken, NJ: Wiley, 2002.

Casarett, L. J., J. Doull, and C. D. Klaassen, eds. *Casarett and Doull's Toxicology: The Basic Science of Poisons*. 6th ed. New York: McGraw-Hill, 2001.

Dixon, W. J. *Design and Analysis of Quantal Dose-Response Experiments*. Los Angeles: Dixon Statistical Associates, 1991.

Marek Brabec

Life Expectancy

Category: Medicine and Health.
Fields of Study: Algebra; Data Analysis and Probability.
Summary: Estimating life expectancy in present populations relies on actuarial tables.

Life expectancy for an individual is the average number of years remaining until death. It is often used to quantify risk of certain characteristics or behaviors as well as to evaluate and compare populations in terms of economics and health. For example, in the United States the life expectancy for a single female currently age 35 is 50.1 years using the 2010 Social Security mortality table. Life expectancy can also be applied to machines or appliances, for product development, to manufacturing quality control, and for the determination of warranty periods. Most incandescent light bulb packages have the life expectancy printed on the packaging. A typical value is 900 hours of use. In this type of application, life expectancy is used as a measure of quality. The calculation of life expectancies can be as simple as taking averages, but normally it uses more advanced mathematics or sampling.

Human Life Expectancy

For human populations, factors affecting life expectancy include resource availability, sanitary practices, healthcare quality, war and sociopolitical factors, cultural and behavior factors, genetic and demographic factors, environmental factors, and epidemics. An increase or decrease in life expectancy may be quoted to describe the risk of a behavior or activity. As an example of using mathematics to make decisions, mathematician James Stein provides the statistic that each hour driven on an interstate highway decreases life expectancy by 19 minutes, while each hour flying decreases life expectancy by only 13 minutes, thus illustrating that flying may be a safer mode of transportation. To quantify the risk in smoking, the U.S. Centers for Disease Control and Prevention (CDC) states that the average life expectancy for a smoker is approximately 14 years less than for a nonsmoker.

Comparing Populations

The life expectancy of newborns is often quoted to compare the relative health of populations in different geographic areas as well as for differences between ethnic or socioeconomic groups, sexes, historical periods, or age groups. The populations being compared may differ in time, geographic region, or demographic characteristics. To compare populations from different time periods, the life expectancy of a newborn in the United States in the early 1900s was about 47 years, improving to about 60 years by the mid-1930s, and further improving to about 78 years by 2009. Life expectancy can vary by gender and race. Historically, females have typically exhibited a higher life expectancy than males. The life expectancy of a newborn female in the United States was estimated to be 80.2 years in 2006, compared to just 75.1 years for a newborn male. Also in 2006, a newborn white male had a life expectancy of 75.7 years, compared to 69.7 years for a newborn black male. According to the United Nations World Population Prospects 2006 Revision, the world life expectancy for a newborn in 2005–2010 is estimated to be 67.2 years, with Swaziland exhibiting the lowest life expectancy at birth for an individual country—approximately 40 years. The latter is often attributed to the high HIV/AIDS mortality and poor healthcare and socioeconomic conditions in sub-Saharan Africa.

For populations that lived in the past, the life expectancy can be calculated by taking the average of the age at death for all of the individuals who lived in the population of interest. For this type of calculation, one normally needs detailed records of dates of births and deaths for the entire population. The first life tables constructed in this way are attributed to John Graunt (1620–1674), who also provided estimated life expectancies in his tables. Following Graunt, a notable life table constructed from birth and funeral data for the purpose of determining life annuity values was published in 1693 by Edmund Halley (Halley's Comet is named after him) for the city of Breslaw, Poland.

Halley used this city for his table because he thought Breslaw was representative of an average European population at the time. Interestingly, Halley provided his own definition of "life expectancy" in describing the third use of his table. In Halley's description, the expected future years a person of a certain age can reasonably expect to live is the proposed number of years upon which an even wager, which is a bet with a 50-50 chance of being won, can be made that the person arrives at that age before he dies. Halley's description is that of the median future lifetime, which differs mathematically from the more modern definition of life expectancy.

Sampling and Estimation

In the absence of complete data, modern statistical methods, including sampling, are used to estimate the average age at death. Similar statistical methods are used to estimate the life expectancy of appliances, components, and machines. In the case of inanimate objects, life expectancy may be interpreted as the average time to failure. To estimate the average time to failure, a sample may be taken and tested in a laboratory environment, or failure statistics may be kept after the product goes to market. The failure rates obtained from such data not only provide a basis for determining the life expectancy of the product, but also can be used in determining the cost of a warranty or guarantee issued by the manufacturer.

In modern populations, actuarial tables are developed that estimate the probability of death at any particular age. These probabilities are used to calculate the life expectancy for an individual at his or her current age. For example, suppose a male age 96 is within a population whose mortality table indicates the probability of a male age 96 dying before age 97 is 0.45; the probability of surviving to age 97 and dying before age 98 is 0.35; and the probability of surviving to age 98 and dying before age 99 is 0.2. Then the expected age at death is calculated as the expected value,

$$96(0.45) + 97(0.35) + 98(0.1) + \frac{1}{2} = 97.25.$$

Hence, the life expectancy is 1.25 years. The term "1/2" in the expected age at death calculation reflects the assumption that the individual dying within the year lives on average one-half the year.

Further Reading

Centers for Disease Control and Prevention. "Annual Smoking-Attributable Mortality, Years of Potential Life Lost, and Productivity Losses-United States, 1997–2001" 54, no. 25 (2005). http://www.cdc.gov/tobacco/data_statistics/mmwrs/byyear/2005/mm5425a1/highlights.htm.

Halley, Edmund. "An Estimate of the Degrees of the Mortality of Mankind." *Philosophical Transaction* 196 (1692). http://www.pierre-marteau.com/editions/1693-mortality.html.

Stein, James. *How Math Can Save Your Life (And Make You Rich, Help You Find The One, and Avert Catastrophes)*. Hoboken, NJ: Wiley, 2010.

United Nations. Department of Economic and Social Affairs, Population Division. "World Population Prospects: The 2006 Revision" (2007).

U.S. National Center for Health Statistics. National Vital Statistics Reports (NVSR) "Deaths: Final Data for 2006" 57, no. 14 (2009).

Kevin L. Shirley

Light

Category: Weather, Nature, and Environment.
Fields of Study: Algebra; Representations.
Summary: Now understood as both a particle and a wave, light is a recurring subject of interest in physics.

Light, a form of electromagnetic energy, mediates the electrostatic interactions between particles. Under some experimental conditions, it acts as a particle, and under others, as a wave. Attempts by physicists to reconcile this dual nature and to otherwise exploit this duality have been the impetus for the development of large areas of mathematics.

Particle or Wave?

Isaac Newton advocated the particle nature of light, initiating the study of geometric, or ray, optics. This form of optics treats light as rays that travel in straight lines, though capable of bending near objects. It is based on two laws. The law of reflection states that when light is reflected from a surface, the angle of incidence equals the angle of reflection. The law of refraction says that light will bend when it passes from one medium to another according to Snell's Law, named for mathematician Willebrord Snell, a relation between the angles of incidence and refraction and light's speed in the two media.

At about the same time, Christiaan Huygens discovered polarized light and explained it with a wave theory. From this beginning, Thomas Young and Augustin-Jean Fresnel developed physical optics. The resulting mathematics allowed engineers to construct extremely faithful lenses; its close cousin, wave acoustics, helped architects design performance halls. Scientists pursued these optics to ever-finer scales. Eventually, they developed the electron microscope, which permits biologists to see individual DNA molecules. Physical biochemists use a related technique called "crystallography." When X-rays are shot through crystals of protein molecules, they form a diffraction pattern, which when transformed by a technique called Fourier analysis (named for mathematician and physicist Joseph Fourier) allows the precise determination of the protein's atomic structure. Many owe their Nobel Prizes to this transformation.

The wave theory of light provides the most natural explanation for the spectrum of visible light. What the physicist calls "light" varies from about 10^{23} cycles per second, corresponding to gamma rays, down to roughly 1000 cycles per second for the electron waves in plasma. What humans can see is but a small part of this, varying from purple at a wavelength of 380 nanometers (nm) or 7.8×10^{14} cycles per second, to red at about 780 nm, or 3.8×10^{14} cycles per second.

Light Speed

In 1861, James Clerk Maxwell wrote down his famous equations describing the interactions between electric and magnetic fields in terms of their sources. Four years later, he derived from them an electromagnetic wave equation, which physicists soon understood to be a description of light waves. In 1907, Edward Rosa and Noah Dorsey used these equations to calculate the speed of light at 299,784 km/sec. The accuracy of this calculation was not matched by experiment until 1926, when Albert Michelson obtained a value of 299,796 km/sec. In 1983, the 17th Conférence Général des Poids et Mesures established a new standard for the length of the meter by fixing the speed of light at 299,792,458 meters/second.

In the 1890s, Hendrik Lorentz, George Fitzgerald, and Joseph Larmor noticed that Maxwell's equations did not change under a certain type of transformation. Henri Poincaré called these "Lorentz transformations" and noticed that they formed a group of symmetries on four-dimensional space-time. Albert Einstein incorporated this symmetry into his theory of special relativity. One of the key postulates is that light travels at the universe's speed limit and so nothing can travel faster. Hermann Minkowski developed from these theories a four-dimensional geometry called "Minkowski space," in which Einstein's famous theory is understood as geometric properties of the space.

Quantum Phenomena

Light held yet further mysteries. Nineteenth-century physics predicted that heated bodies should radiate

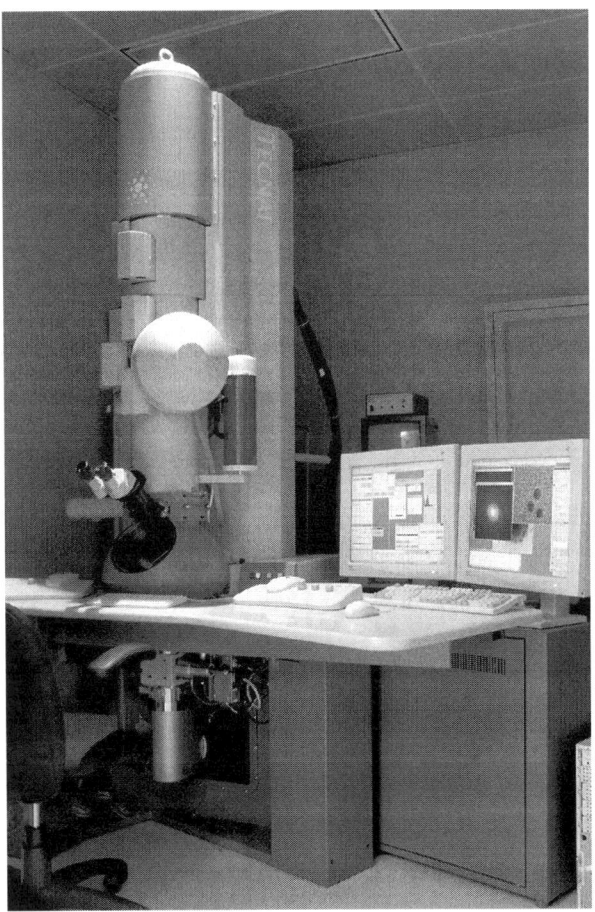

The National Center for Electron Microscopy houses this microscope with a 2048x2048 CCD camera. (National Center for Electron Microscopy, John Turner)

infinite amounts of energy and that an atomic electron should plunge into the nucleus. Max Planck eliminated the first problem by postulating the quantization of light. Einstein used this idea to explain properties of the photoelectric effect, the phenomenon behind solar panels. Niels Bohr expanded these ideas into an explanation of why electrons in atoms do not continuously radiate light until they collapse into the nucleus. All three won Nobel Prizes for their work and quantum physics was born.

John von Neumann developed a mathematical description of these quantum phenomena involving Hilbert spaces and operator algebras. As a result, research into Operator Algebras became a major research focus of the last half of the twentieth century. To further explain quantum behavior, von Neumann and Garrett Birkoff developed quantum logic, a subject pursued not only by mathematicians but also by many philosophers. In a high point of this endeavor, John Bell developed the Bell inequalities in 1966. Sixteen years later, Alain Aspect confirmed that quantum systems do violate these inequalities, and provided strong evidence that the mysterious results of quantum mechanics are not solely because of our difficulties in measuring systems on such a fine scale but are because of the very nature of these small-scale systems. These experiments exploited a quantum property called "entanglement." Richard Feynman hypothesized this entanglement might be exploitable as a computational resource. In recent decades, Peter Schor, Lov Grover, and others have developed algorithms based on Feynman's idea and created the field of quantum computing.

Quantum mechanics has, in the last half century, developed into quantum field theory (QFT). QFT attempts to explain all particles and forces by equations that are modeled on Maxwell's. In developing their models, mathematical physicists rely on physical properties to perform manipulations mathematicians find objectionable because of their lack of rigor. Many great mathematicians have taken up the challenge of developing a rigorous axiomatic basis for QFT. Lying at the intersection of philosophy, mathematics, and physics, many mathematicians see this as one of the great challenges of the twenty-first century.

Further Reading
Baierlein, Ralph. *Newton to Einstein: The Trail of Light*. Cambridge, England: Cambridge University Press, 1992.
Farndon, John. *From Newton's Rainbow to Frozen Light: Discovering Light*. Chicago: Heinemann Library, 2007.
Smith, Francis Graham, and John Hunter Thomson. *Optics*. Hoboken, NJ: Wiley, 1971.
Sobel, Michael. *Light*. Chicago: University of Chicago Press, 1987.
Zeilinger, Anton. *Dance of the Photons: From Einstein to Quantum Teleportation*. New York: Farrar, Straus & Giroux, 2010.

MICHAEL KLUCZNIK

Lightning

Category: Weather, Nature, and Environment.
Fields of Study: Data Analysis and Probability; Geometry; Measurement; Representations.
Summary: Lightning is studied, modeled, and predicted using mathematical techniques.

Lightning is an electrical phenomenon of nature that has been observed by people around the world for thousands of years. Thunder is the sound of lightning, created by the intense heat of a lightning bolt. Many people may have learned as children a simple calculation for estimating the distance of lightning based on the sound of thunder. Since thunder travels about one mile in five seconds, a 15-second delay between the time lightning is seen and the time the thunder is heard indicates that the lightning strike was about three miles away.

Lightning strikes occur frequently around the globe, with an estimated 25 million cloud-to-ground strikes per year in the United States alone. Lightning has a large number of religious associations, and it is often used as a metaphor for sudden insight or inspiration. Mathematician Carl Friedrich Gauss is reported to have said, regarding a problem he had been working on, "Like a sudden flash of lightning, the riddle was solved." Lightning is studied by mathematicians, often in collaboration with scientists in other fields, to better understand the various facets of this complex phenomenon.

Among the several types of lightning that occur, the most commonly seen and the most dangerous is cloud-to-ground lightning, caused by the discharge of electrons into the Earth from thunderclouds in the atmosphere. The voltage released by a bolt of cloud-to-ground lightning is on the order of 1 million times the voltage in a standard electrical outlet.

The excess of electrons at the base of a thundercloud repels electrons on the ground deep into the Earth, inducing a strong positive charge on the ground below. While air usually acts as an insulator, preventing the flow of electric current, the strong electric field between a storm cloud and the Earth can reach tens of thousands of volts per inch, pulling air molecules apart into negatively charged electrons and positive ions. This creates pathways of ionized air known as "streamers." The freely moving charges in the ionized air allow electric current to flow through it.

A lightning strike occurs when a streamer carrying electrons from the cloud toward the Earth meets a shorter, positively charged airstream reaching up from an object on the Earth. This creates a complete conductive pathway between the cloud and the ground and a sudden and massive discharge of electrons into the Earth.

Between an average thundercloud and the Earth, there are an estimated 10^8 volts, reaching 10^9 (1 billion) volts in more-intense strikes. For perspective, one may compare 1.2×10^8 volts between a thundercloud and the Earth to the 120 volts delivered by a standard electrical outlet in the United States:

$$\text{Voltage between cloud and ground}$$
$$= 1.2 \times 10^8 \text{ volts}$$
$$= 1.2 \times 10^2 \times 10^6 \text{ volts}$$
$$= 120 \text{ volts} \times 10^6$$
$$= \text{Voltage in standard electrical outlet} \times 1 \text{ million.}$$

Lightning has been observed around the world for thousands of years. It has been spotted not only during thunderstorms but in volcanic eruptions, intense forest fires, heavy snowstorms, and large hurricanes. (Photos.com)

The heat created by the electric current in a bolt of lightning reaches temperatures up to 30,000 kelvins (K), more than five times the temperature of the surface of the sun and hot enough to melt rock and fuse soil and sand into glass. The temperature on the Kelvin scale is the temperature in degrees Celsius plus 273.15. The intense heat in a channel of lightning causes the air within the channel to expand rapidly, sending out a shock wave that weakens into the acoustic wave of thunder. The electric current and heat of a lightning strike can start forest fires, damage a property, destroy electrical equipment, and cause serious or fatal injuries to people and animals. According to estimates by the National Weather Service, lightning causes on average about 60 deaths and 300 injuries in the United States each year.

Statistics collected by NASA satellites have found that most of the eastern half of the United States sustains about eight flashes of lightning per square mile per year (decreasing to less than one per square mile per year toward the West Coast). Since 1 mi^2 = 640 acres, this translates to eight flashes per 640 acres per year, or one flash per 80 acres per year. Accordingly, a one-acre lot in this region would be struck by lightning on average once every 80 years.

Mathematical research can help to predict the behavior of lightning strikes based on weather patterns and other variables; for example, by modeling probabilistic distributions of lightning strikes according factors such as time, geography, and strength. The mathematical theory of highly optimized tolerance (HOT) is useful in controlling forest fires caused by lightning. This theory suggests optimal placement of fire breaks: if data or other evidence suggests that lightning strikes some areas of a forest more frequently than others, then large fires can best be prevented by purposefully cutting fire breaks that create sections whose sizes are inversely proportional to the rate at which lightning strikes. Other mathematicians are interested in studying the patterns and geometry of lightning. Mathematician Benoit Mandelbrot, known for his study of fractal patterns, noted that lightning does not travel in a straight line but rather in patterns reminiscent of fractals. Techniques of fractal modeling are used to study fractal patterns in the ionized plasma structures of lightning streamers. Morphological filtering and gradient detection can be used to help visualize lightning in satellite imagery and separate it from other visible effects, such as city lights.

Further Reading

Krider, E. Philip. "Lightning Damage and Lightning Protection." In *The Thunderstorm in Human Affairs*. 2nd ed. Edited by Edwin Kessler. Norman: University of Oklahoma Press, 1988.

Rakov, Vladimir, and Martin Uman. *Lightning: Physics and Effects*. Cambridge, England: Cambridge University Press, 2007.

Uman, M. A. *Lightning*. New York: Dover, 1984.

Barbara A. Shipman

Maps

Category: Travel and Transportation.
Fields of Study: Geometry; Measurement.
Summary: Scales and projections are used to display geographic features on maps.

The word "map" is the name given to any representation of the Earth's features—natural and artificial—usually on a plane using a given scale and map projection. In scientific and mathematics applications, the term "map" is more broadly interpreted. The purpose of a map is to register and transmit information about those features and the spatial relations between them.

A common characteristic of all maps is that they are reduced and conventional representations of reality, which makes them significantly different from an aerial photograph. While an aerial photograph depicts all the physical objects that a sensor could detect and register (and only those), a map is a selection of natural and artificial objects, visible and invisible, chosen to fit the cartographer's purpose and the limits imposed by the available space. These objects are represented on maps in a conventional way by means of symbols; this is not the case with photographs, in which they are depicted by the visual image they present when viewed from above by the sensor. The symbols in a map are designed to categorize features by type and to optimize the document's legibility. Very often, their size is not proportional to the size of the objects they represent. For example, roads are symbolized by lines of variable thickness and pattern, often much larger than the corresponding width of the actual roads, since representing them to exact scale would often make them too thin, even invisible. In other cases, such as with cities, features are symbolized by punctual symbols whose color and shape depend on the classification scheme chosen (such as administrative status or population).

Maps are usually classified in three main categories: general reference maps, thematic maps, and charts. A general reference map depicts generic geographic information of various types considered useful to a large spectrum of users. This information may include topography, political and administrative borders, and land cover. The best example of a general reference map is the topographic map. A thematic map, on the other hand, represents the geographic distribution of a specific theme or group of themes such as geological features, population, or air temperature. A chart is a special type of map designed to support navigation, either maritime (with nautical charts) or aerial (with aerial charts).

History

Maps were first made by the ancient civilizations of Europe and the Middle East several centuries before the Common Era. One of the oldest known is a Babylonian clay map of the world c. 600 B.C.E., now kept in the British Museum. Though it is documented in the testimony of Ptolemy of Alexandria (c. 90–169 C.E.) and others that maps were drawn in Greece as early as the seventh century B.C.E., none are known to have survived. However, several medieval manuscript maps have survived that represent the ecumene (the known inhabited part of the world around the Mediterranean basin). Few had any practical purpose, and most were symbolic representations inspired by religion and myth rather than by reality. In his *Geography*, published for the first time in the second century C.E., Ptolemy describes three map projections in detail and presents a list of more than 8000 places in the ecumene, defined by their latitudes and longitudes.

This list permitted others to redraw the maps that may have accompanied the original text once the work was translated into Latin and disseminated throughout Europe during the fifteenth century. The publication of several editions of *Geography* did much to bring about the rebirth of scientific cartography. By this time, nautical charts had already been used to navigate in the Mediterranean for at least two centuries. And while terrestrial cartography quickly adopted the geographic coordinates and map projections proposed by Ptolemy, nautical charts remained based on the magnetic directions and estimated distances observed by pilots at sea. Still, these representations were of astonishing accuracy and detail compared with the traditional maps of the time.

It is now known that the first nautical charts, commonly known as "portolan charts," were constructed in the first half of the thirteenth century, probably in Genoa, after the introduction of the magnetic compass and the adoption of the decimal system in Europe. This basic model continued to be used in nautical cartography for a long time, though much improved by the introduction of astronomical navigation during the fifteenth century. The resulting modality, based on observed latitudes and magnetic directions, became known as the "latitude chart" (or "plane chart") and played a fundamental role in the discoveries and maritime expansion periods. In 1569, an important world map specifically conceived for supporting maritime navigation was constructed by the Flemish cartographer Gerard Kremer (1512–1594), better known by the Latinized name of "Gerardus Mercator." Contrary to traditional portolan charts, this map was based on the latitudes and longitudes of places and represented all rhumb lines (lines of constant course) as straight segments making true angles with the meridians.

Though Mercator did not explain how the planisphere was made, a geometric method was most likely used. The mathematics of the projection is not trivial and its formalization had to wait until after calculus was developed, more than one century later. As for its full adoption as a navigational tool, that did not occur until the middle of the eighteenth century, when the marine chronometer was invented and longitudes could finally be determined at sea.

Mathematical Cartography

Maps may depict only a small part of the whole surface of the Earth. The word "scale" means the quotient between a length measured on a map and the corresponding distance measured on the Earth's surface. Because it is not possible to represent the spherical surface of the Earth in a plane without distorting the relative position of the places (and thus, the shape of all objects), the scale of a map is not constant, always varying from place to place and, in the generality of cases, also with the direction. In large-scale maps, like the plant of a city or the topographic map of a small region, these distortions can be ignored and the scale considered constant for most practical purposes. That is not the case when a large area of the Earth's surface is represented, like in a planisphere or a map of a whole continent. Here shapes may be strongly deformed and the scale varies significantly from place to place. Measurements made on those maps with the purpose of evaluating distances between places, using their graphical or numerical scale, are only approximations, as the scale strictly applies only to certain parts of the maps (like the central meridian or parallel), and their use in the other regions may lead to very large errors.

"Map projection" refers to any systematic way of representing the surface of the Earth on a plane. The process consists of two independent steps. First, one has to replace the irregular topographic surface, with all its mountains and valleys, with a simpler geometrical model, usually a sphere or an ellipsoid where a system of geographic coordinates (latitude and longitude) is established. Second, one has to project that model onto a plane surface. This step may be accomplished by some geometric construction or by a mathematical function that transforms each pair of geographic coordinates latitude (j) and longitude (l) into a pair of Cartesian coordinates x and y, defined on the plane. Depending on the purpose of the map, there are many different map projections to choose from. Knowing that none of them conserves the relative position of all places on the surface of the Earth, the choice is usually driven by the type of geometric property one wants to preserve. For example, equivalent or equal-area projections conserve the relative areas of all objects and are typically used in political maps. Conformal projections conserve the angles around any point on the map (the scale does not vary with direction), as well as the shape of small objects, and are utilized in nautical charts and topographic maps. Equidistant projections conserve the scale of certain lines and are used whenever one wants to preserve distances measured along those lines. This is the case of the azimuthal equidistant projection, where distances measured from the center of the projection along all great circles are conserved. This property is useful, for example, for quickly determining the distance of any place in the world measured from a chosen location.

However, it is not possible for a map projection to have all these properties at the same time, and the conservation of some properties is usually accompanied by significant distortions of the others. A significant example is the Mercator projection (which is conformal), where all rhumb lines are represented by straight segments making true angles with the meridians. However, the scale increases with latitude in this projection, strongly affecting the proportion of the areas. The branch of cartography dealing with map projections is known as "mathematical cartography." Though some map projections have been well known since remote antiquity, when they were often used for representing the sky, a more formal approach became possible only after the development of calculus. The most important contributions in the formalization of mathematical cartography were those of Johann Heinrich Lambert (1728–1777), Joseph-Louis Lagrange (1736–1813), Carl Friedrich Gauss (1777–1855) and Nicolas Auguste Tissot (1824–1897).

Computers and geographic information systems have made it possible for previously unforeseen numbers of users to produce good-quality maps tailored to their specific needs and at a reasonable cost. They also allow scientists and mathematicians to map increasingly complex systems and concepts, such as the universe and the World Wide Web. They can also often render in three dimensions and beyond. In mathematics, maps can be used to alternatively express functions or connect mathematical objects. In conceiving those systems, as well as in acquiring the geographic data necessary to construct the representations within, mathematics continues to play a fundamental role.

Further Reading

Brown, Lloyd A. *The Story of Maps*. New York: Dover Publications, 1977.

Bugayevskiy, Lev, and John Snyder. *Map Projections. A Reference Manual*. Oxfordshire, England: Taylor & Francis, 1995.

Ehrenberg, Ralph E. *Mapping the World: An Illustrated History of Cartography*. Washington, DC: National Geographic, 2005.

Snyder, John. *Flattening the Earth. Two Thousand Years of Map Projections*. Chicago: University of Chicago Press, 1993.

Zuravicky, Orli. *Map Math: Learning about Latitude and Longitude Using Coordinate Systems*. New York: Rosen Publishing, 2005.

Joaquim Alves Gaspar

Molecular Structure

Category: Medicine and Health.
Fields of Study: Algebra; Geometry; Representations.
Summary: The geometry of molecules can be an important property, as in the shape of a protein molecule or the double-helix of DNA.

The physical structure of molecules is important in chemistry, biology, physics, and engineering. The precise structure can influence the chemical reactivity of a molecule as well as its response to other physical interactions, such as how it can absorb energy in the form of photons (light particles or X-ray particles). These interactions can have important implications in biology, medicine, health, and engineering. For instance, how proteins fold determines their function, and the shapes of certain protein molecules influence the existence of diseases. For example, shape is important in the normal function of the hemoglobin molecule, the molecule crucial for absorbing oxygen in red blood cells so that they can transport it throughout the body.

Hemoglobin consists of four protein subunits, associated with four heme subunits (ring-like structures containing an iron atom). As one oxygen molecule (O_2) binds to one of the heme units, the molecule distorts so as to allow another oxygen molecule to more readily bind in a cooperative way to another heme unit. This in turn distorts the molecule so that another O_2 finds it even more readily. Altogether, four O_2 molecules can ordinarily bind to one hemoglobin molecule. In sickle-cell anemia, two mutations in two of the four protein units distort the hemoglobin molecule so that the misshapen units form long chains. These in turn cause the red blood cell to become misshapen and lose its elasticity so that it can no longer readily move through small capillaries. Besides being painful, the misshapen red blood cells are destroyed by the spleen, resulting in anemia.

Another example of how the shape of a protein can cause disease is that of prions, which are misshapen proteins that enter (or "infect") cells and cause the cells' proteins to become misshapen. Prions are probably best known as the cause of bovine spongiform encephalopathy (commonly called "mad cow disease") in cattle. Finally, protein folding is also implicated in Alzheimer's disease. Thus, there is natural interest in understanding how these molecules fold. Knowing

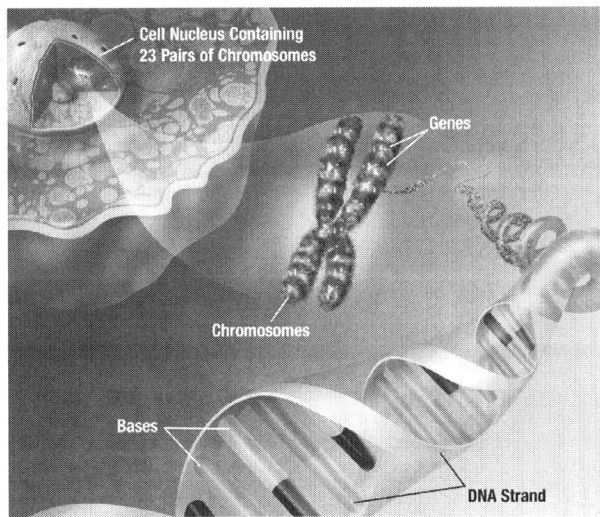

An illustration created for a National Institutes of Health study on DNA and Alzheimer's disease. (National Institutes of Health)

precisely how any particular protein folds in a particular chemical environment generally requires intensive mathematical computations that implement various equations from the area of physics known as "quantum mechanics." It is interesting that while supercomputers are usually used for this work, dozens of scientific articles have been written that instead relied on computations performed by harnessing millions of ordinary PCs, volunteered by millions of individuals—over 5 million CPUs as of September 2010.

DNA

Besides proteins, another important molecule studied extensively for its structure is DNA. While the double helix structure has been known for over 50 years, precisely how DNA is used in the cells of the body is still a source of research in the twenty-first century. In order to fit inside a cell nucleus, DNA must be very tightly coiled. How the appropriate sequence of DNA that a cell might need at a particular time can be rapidly located and then rapidly transcribed into messenger RNA for making a particular enzyme of interest is a complex process. Simply understanding how unknotting the knotted DNA takes place within the nucleus is nontrivial, and the mathematical discipline known as "topology" (and its subdiscipline, knot theory) has helped to elucidate how the cell handles the knotted DNA. One key equation to help understand the process of DNA supercoiling is $Lk = Tw + Wr$, where Lk is the linking number, Tw is the twist, and Wr is the writhe. This equation, attributed to G. Calugareanu, J. H. White, and F. B. Fuller, relates the linking number of the DNA (which essentially describes how the two backbones of the double-stranded DNA are linked) to the twist (the twisting of either backbone relative to the central axis of the DNA) and the writhe (which relates how the central axis of the DNA is oriented in three-dimensional space).

Other Structures

Besides proteins and DNA, molecular modeling is important in other areas. In the past, a scientist looking for a chemical that would have a certain effect in a certain situation, given a compound that reacts in a slightly different way in a slightly different situation, would likely have changed one part of the molecule and tested the new product; changed another feature and tested that product; and so on. Combinatorial chemistry is devoted to trying to automate the synthesis—and efficacy studies—of a huge number of different permutations of some basic chemical structure, somewhat in parallel. Interest in combinatorial chemistry is widespread among pharmaceutical companies.

Determining the molecular structure of molecules often relies on the general area of spectroscopy, which involves examining the spectrum that results when visible, ultraviolet, or infrared light or X-ray radiation, is applied to molecules. Mathematics that can categorize the different types of symmetry that molecules can assume can be used to help spectroscopy determine what shape the molecule must have. As one example, analysis of DNA in crystalline form by X-ray crystallography led to James D. Watson and Francis Crick's determination of the double-helix structure of DNA in 1953.

More recently, a form of pure carbon was found to be created from an electric arc between graphite electrodes (or from high-temperature burning of gaseous hydrocarbons). The carbon compounds created are known as "fullerenes," which are cage-like in appearance. The first fullerene to be discovered and have the results scientifically published is now known as Buckministerfullerene or C_{60}. Discovered in 1985 by Richard Buckminster "Bucky" Fuller, it was determined to essentially look like a soccer ball in appearance (a truncated icosahedron). How the precise polyhedral cagelike structure was determined from spectroscopy relied heavily on mathematics, specifically the area of abstract algebra known as "group theory," applied to quantum mechanics. Whereas fullerenes like C_{60} and C_{70} are cage-like, other pure forms of carbon obtained from graphite that do not fully close up include nanotubes. While fullerenes and nanotubes may have health applications, they are also of interest purely as nanotechnological objects. Indeed, some nanotubes are extremely strong and one day may make superstrong fibers; some, when other atoms such as potassium are added, are superconductors. For instance, the orientation of carbon atoms in nanotubes affects electrical conductivity (whether the molecules are conducting or semiconducting).

Another approach to determining molecular structure, particularly to surfaces, is to use instrumentation such as the scanning tunneling microscope. This tool relies heavily on physics (quantum mechanical tunneling) principles.

Further Reading

Plunkett, Mathew J., and Jonathan A. Ellman. "Combinatorial Chemistry and New Drugs." *Scientific American* 276 (1997).

Schlick, Tamar. *Molecular Modeling and Simulation: An Interdisciplinary Guide*. New York: Springer Verlag, 2006.

Sumners, De Witt. "Lifting the Curtain: Using Topology to Probe the Hidden Action of Enzymes." *Notices of the American Mathematical Society* 42 (1995).

Rick Kreminski

Moon

Category: Space, Time, and Distance.
Fields of Study: Data Analysis and Probability; Geometry.
Summary: Though mankind has always looked up at the moon and even visited, most of the body of lunar knowledge is actually contributed by mathematics, which continues to attempt to model its motion.

The moon is the sole natural satellite of the Earth. Specific astronomical searches have established positively that the Earth has no other satellites larger than a few meters. The lunar body is nearly a sphere with a mean radius of 1738 kilometers (km) or 1000 miles—only 3.7 times less than the Earth. The mean distance of the moon from the Earth is 384,400 km (238,855 miles). The moon is the fifth largest satellite in the solar system and the largest one relative to the size of its planet. The moon is so near and so large in comparison with its "host" that the entire system is often dubbed the "double planet."

Viewed from above the North Pole of the Earth, the moon travels around it counterclockwise in a slightly elliptical path. The sideric month (one orbit around the Earth with respect to the stars) is 27.3217 days. The synodic month (the cycle of phases visible from the Earth; for example, the time interval between two successive "new moon phases") is 29.5306 days.

The period of one spin of the moon around its axis (a "lunar day") is exactly equal to the sideric month because of tidal breaking. This phenomenon is also known as "synchronous rotation," or tidal coupling." As a result, from the Earth, people can observe only half of the lunar surface (called the "near," or "visible," "side"). The "far" (called "invisible") hemisphere was photographed for the first time in 1959 by the Soviet robotic spacecraft *Luna-3*, an episode of the space race between the United States and the Soviet Union. On the moon, the disk of the Earth does not rise and set. It is observable only from the near side in an almost permanent point of the lunar sky (fluctuating a little from a small phenomenon called "libration").

The face of the moon was influenced by both internal and external factors. On the surface, observers distinguish so-called darker "maria" (flat "seas" without water) and brighter highlands. All of them are covered with numerous craters, the highlands more so than the seas. The far side of the moon has practically no seas. Because of constant bombardment by various small interplanetary particles, the entire surface is enveloped with thin fractured material called "regolith." There is no atmosphere on the moon. As a result, the difference in temperatures between a lunar day and a lunar night is very high: between –170 degrees Celsius and +130 degrees Celsius (–274 degrees Fahrenheit to +266 degrees Fahrenheit). Water in the form of subsurface ice exists in polar regions. There are no traces of modern tectonics on the surface.

From the Earth, the visible angular diameter of the moon is 0.5 degrees and fairly close to the angular diameter of the sun. This property is essential because sometimes the three bodies, the sun, the Earth and the moon, align along a straight line. In this case, humans observe either a total lunar (if the moon is farther from the sun than the Earth) or a total solar (if the moon is between the sun and the Earth) eclipse. The latter is visible only within narrow strips on the Earth. Such observations are important for solar physics. To see these phenomena, astronomers regularly organize special expeditions. Eclipses often held great religious significance. Scholar Anaxagoras of Clazomenae explained the phenomenon using mathematics. He was imprisoned for asserting that the sun was not a god and that the moon reflected the sun's light.

The age of the moon is about 4.5 billion years, which is close to the age of the sun and the entire solar system. Of the various concepts of the moon's origin, the prevailing hypothesis is that the Earth-moon system was formed by a giant impact: a planet-sized body hit the nearly formed proto-Earth, ejecting material into orbit around the proto-Earth, which accreted to form the moon.

The mean density of the moon is just 3.34 grams per centimeter^3 and, as a result, the mass of the moon is 81 times less than that of the Earth. The interior of the moon is geochemically differentiated: it has a distinct crust, mantle and core. Surface gravity on the moon is six times less than on the Earth. The general magnetic field of the moon is practically absent.

The moon has always played a significant role in religion, science, art, and culture. Since the Paleolithic, the lunar orb in the sky has been utilized for calendar purposes. That is why the similarity of the terms "moon" and "month" is not coincidental. For the philosopher Aristotle, the moon marked a great border between a mortal and corruptible sublunar (terrestrial) world and an immortal world of ideal heavenly bodies. It became a significant symbol for Islam. For Isaac Newton, the moon was the prime test body to demonstrate mathematically that the fall of an apple and the orbiting of a celestial body are ruled by a single natural law of universal gravity.

Mathematical Modeling

Many mathematicians have developed theoretical models for the motion of the moon. The exact path of the moon around the Earth is affected by many perturbations and is extremely complicated. That is why, after Newton, research of lunar motion (lunar theory) became the central problem of celestial mechanics. Consequently, it appeared among the most critical and difficult tasks for applied mathematics. The moon's gravitational influence on the Earth produces the ocean tides and the tiny lengthening of the calendar year. Most of what we know about the moon's size, shape, and other properties has been derived largely through mathematical computations, using mathematical theory and data from Earth-based observations, satellite imagery, and direct measurements made by astronauts.

Human Exploration

Starting at least from Roman times, science fiction authors were the forerunners for delivering terrestrials to the moon. In reality, the first space robots to the moon were launched by the Soviets in 1959. But they failed in the space race with the United States to realize manned expeditions. The first terrestrials to visit the moon were the American astronauts of the Apollo program. After preliminary robotic programs (Ranger, Lunar Orbiter, and Surveyor) and Apollo flybys, American manned landings on the moon occurred in 1969–1972. Among seven planned landings (from Apollo-11 up to Apollo-17), six missions were tremendously successful. Twelve crewmembers stepped down on the near side of the moon, and six more orbited it. Astronauts performed a number of experiments and returned to the labs about 382 kg of lunar matter. Since 2004, Japan, China, India, the United States, and the European Space Agency have each sent successful automatic lunar orbiters.

Among the many thousands of contributors to lunar programs, mathematicians often played outstanding roles. One significant individual was mathematician Richard Arenstorf, who solved a special case of the three-body problem with figure-eight trajectories now called "Arenstorf periodic orbits." In 1966, he was awarded a NASA medal for exceptional scientific achievement for this work. Another was Evelyn Boyd Granville, who used numerical analysis to aid in the design of missile fuses. She later worked on trajectory and orbit analyses for several space missions, including Apollo. She said, "I can say without a doubt that this was the most interesting job of my lifetime—to be a

Astronaut Harrison H. Schmitt standing next to a huge boulder during the Apollo 17 mission to the moon. (National Aeronautics and Space Administration)

member of a group responsible for writing computer programs to track the paths of vehicles in space." In fact, mathematicians occupied many seats in the first row of the Mission Control center. Their work was critical for calculating trajectories and for maneuvers that involved the meeting of two objects in space, including landing on the moon. They also played a significant role in determining a rapid and feasible solution that would safely return the damaged Apollo 13 manned spacecraft to Earth.

Among mathematicians in Russia, the most noticeable contribution to flights to the moon was made by Efraim L. Akim of the Keldysh Institute for Applied Mathematics at the Russian Academy of Sciences in Moscow. He was the principal investigator for special lunar orbiters to create a mathematical model of the lunar gravitational field and the leader of a team to calculate trajectories of the Russian lunar robotic spacecraft.

Several international treaties regulate mutual relations of various states with respect to modern space explorations of the moon. The most important among them are the Outer Space Treaty (1967) and the Agreement Governing the Activities of States on the Moon and Other Celestial Bodies (1979).

Further Reading

Gass, S. I. "Project Mercury's Man-in-Apace Real-Time Computer System: 'You Have a Go, at Least Seven Orbits.'" *Annals of the History of Computing, IEEE* 21, no. 4 (1999).

Eckart, Peter, ed. *The Lunar Base Handbook*. New York: McGraw-Hill, 1999.

Stroud, Rick. *The Book of the Moon*. New York: Walker and Co., 2009.

Ulivi, Paolo, and David Harland. *Lunar Exploration: Human Pioneers and Robotic Surveyors*. Chichester, England: Praxis Publishing, 2004.

Alexander A. Gurshtein

Nanotechnology

Category: Architecture and Engineering.
Fields of Study: Geometry; Measurement; Number and Operations.
Summary: Nanoscience relies on mathematical modeling to predict the behavior of substances at the nanoscale.

Nanotechnology is a relatively new field of scientific study, the conceptual origins of which are typically credited to a presentation by physicist Richard Feynman in the late 1950s. A nanometer is one-billionth of a meter, and nanoscience focuses on matter with dimensions between 1 and 100 nanometers. For comparison, an ordinary sheet of paper is about 100,000 nanometers thick, a human hair is between 60 and 120 nanometers thick, and the diameter of one atom of gold is about 1/3 of a nanometer. Thus, nanotechnology is concerned with studying materials at a very small scale, ranging from roughly larger than a single atom at the lower end to objects that can be seen with a high-quality optical microscope at the upper end.

Physicists, mathematicians, and other nanotechnologists are often particularly interested in how the physical, chemical, and biological properties of materials may differ at this scale as opposed to properties of the same materials in bulk or at the scale of single atoms or molecules. Feynman discussed the notion that human beings would someday be able to create increasingly smaller and smaller machines, in part through directed, precision arrangement of atoms and molecules. He also introduced the idea that change in scale would affect the mathematical and physical properties of technology and processes. For example, relatively large-scale forces like gravity would begin to diminish in importance as machines grew smaller, while molecular-level van der Waals attractive forces, named for chemist Johannes van der Waals, and other properties would take on more important roles. However, he did not call his own ideas "nanotechnology." Instead, the first use of the term as it is typically meant in the early twenty-first century is credited to engineer K. Eric Drexler in the 1980s. He also helped spread nanotechnology and molecular manufacturing ideas to a broader audience. There are types of technology that are already being created at nanoscales. Some visions about the future of molecular manufacturing are much like the replicator device in the science fiction franchise *Star Trek*: human-scale or even larger objects, even complex devices like computers, quickly assembled atom by atom.

Any word with the prefix "nano" means at a nanometer scale (for example, the word "nanofilter" would refer to a filter at the nanometer scale), but there are also some basic classifications that are in common use. Nanomaterials are furthered classified as nanoparticles (if all three dimensions are nanosized), nanotubes (which have a nanosized diameter but greater length), and nanofilms or nanosheets (the thickness is nanosized, but the width and height may be much greater). Nanostructured materials have an internal structure that is nanosized, but the pieces of material may be much larger.

Principles

Nanotechnology draws on many scientific fields, including chemistry, physics, and biology, as well as engineering and materials science, and one common thread among them all is mathematics. Interestingly, the extreme difference in size between usual applications and applications at the nanoscale means that some of the most fundamental laws describing natural processes do not apply. For instance Ohm's law describes the flow of electrical current as

$$I = \frac{V}{R}$$

where *I* is the current in amps, *V* is the potential difference in volts, and *R* is the resistance of a conductor in ohms. This law is based on the free flow of electrons and hence does not describe the movement of electrons through nanowires, which may be so narrow as to allow only one electron to pass through at a time. To take another example, at the nanoscale, heat flow is no longer governed by standard continuity boundary conditions and different assumptions that allow for discontinuities must be used instead. Identifying and quantifying how such fundamental laws and expectations change at the nanoscale is one important field of study within nanotechnology.

Construction of systems at the nanoscale allows researchers great control over the form of the nanoparticles developed as well as the ways they form three-dimensional wholes. One line of research involves devising structures that require the minimum number of molecules for a given construct, while another involves developing self-assembling structures, such as cubes and buckyballs. Nanotechnology also adds new complications to issues of dimensionality. From elementary geometry, humans are accustomed to thinking in terms of one dimension (a line), two dimensions (a plane), and three dimensions (a cube, or any object in space). However, at the nanoscale the picture is not so clear. For instance, quantum dots or "artificial atoms" that contain only one or a few electrons with discrete energy states are zero-dimensional solids, which can function in quantum computers as a binary switch. Fractals, which are described by noninteger dimensionality (for example, a two-and-a-half-dimensional object) are also used to model nanoscale systems.

Applications

Medicine is one of the most promising fields for nanotechnology because many internal processes of the human body take place at nanoscale dimensions. Drug delivery is one promising field: nanoparticles can be used to deliver drugs directly to particular cells, for instance, for chemotherapy that targets cancerous cells but not healthy cells and thus reduces tissue damage. Nanotechnology has also developed ways to use nanoshells to concentrate heat from infrared light to

The world's first motorized light-powered nanocars made from only 169 atoms were built at Rice University in 2006. (Rice University, Yasuhiro Shirai)

destroy cancer cells with minimal damage to adjacent healthy cells. Nanotechnology promises to allow some drugs now delivered by injection to be taken orally, encapsulated in a nanoparticle, which would help it pass into the bloodstream from the stomach. Nanofibers have been used to repair damaged joints by stimulating the body's production of cartilage; nanoparticles have been used to increase the speed of blood clotting to prevent blood loss in trauma patients; and nanocrystalline silver is already being used as an antimicrobial agent for wound treatment. Nanocrystal technology is being developed to improve medical imaging, and in the future it may be possible to develop cell repair nanorobots, which could be programmed to repair diseased or damaged cells in a person's body.

Nanotechnology has many applications in the fields of energy production and pollution control. Nanotechnology has made it possible to create more efficient solar cells at a lower cost (making the technology more likely to be adopted) and provided new forms that make solar technology more convenient. For instance, solar cells created by embedding nanoparticles in plastic film can be incorporated into mobile phones and portable computers. Batteries created using nanotechnology can be made lighter and more powerful and can also be charged more quickly than conventional batteries, increasing the efficiency of hybrid automobiles. Nanofilters are increasingly being applied in food production, water filtration, and air pollution control, and nanoparticles are also used in some applications to absorb contaminants.

In manufacturing and construction, nanotechnology has led to the development of new materials that are lighter, stronger, and possess more desirable properties than their conventional analogues. For instance, nanomolecular structures are already being used to make concrete and asphalt more resistant to water, and nanomaterials added to light-emitting diode (LED) lighting makes them more resemble standard lighting, allowing the incorporation of more efficient LED lights in home and industrial use while retaining the look of traditional lighting. Nanocoatings are commercially available that resist corrosion, offer insulation and UV protection, and can remove pollutants from a building's atmosphere.

Further Reading
Foster, Lynn E. *Nanotechnology: Science, Innovation and Opportunity*. Upper Saddle River, NJ: Prentice Hall, 2006.
Garcia-Martinez, Javier, ed. *Nanotechnology for the Energy Challenge*. Weinheim, Germany: Wiley-VCH, 2010.
Matthews, Miccal T., and James M. Hill. "Micro/Nano Thermal Boundary Layer Equations with Slip-Creep-Jump Boundary Conditions." *IMA Journal of Applied Mathematics* 72 (2007).
"Nanotechnology." *Scientific American*. http://www.scientificamerican.com/topic.cfm?id=nanotechnology.
"Nanotechnology News." *Science Daily*. http://www.sciencedaily.com/news/matter_energy/nanotechnology.

Sarah Boslaugh

Nervous System

Category: Medicine and Health.
Fields of Study: Algebra; Number and Operations.
Summary: Mathematicians use a variety of mathematical modeling techniques to map and analyze nervous systems.

Human beings and many animals have two systems that are responsible for regulating and coordinating the activities of the body: the nervous system and the endocrine system. The first provides extremely fast responses, like reacting to touching a hot stove. The second responds more slowly and continuously, such as regulating blood sugar after a meal. Both systems work by detecting internal and external variations, such as shapes, odors, or temperature, to maintain the balance of body functions. Neuroscience, which is the study of the nervous system (including the brain) and its functions, is an interdisciplinary field that draws concepts and methods from many fields such as mathematics, psychology, biology, physics, and medicine. The Hodgkin–Huxley equations, named for Alan Hodgkin and Andrew Huxley, are fundamental to the development of mathematical models and simulations that have long been the basis of many experiments to study the nervous system. Many neuroscience researchers and teachers use the open source NEURON computer simulation system, which incorporates systems of equations and computational algorithms to mathematically model and display the behavior of individual neurons or networks of neurons in a dynamic way that is often difficult or impossible to achieve in traditional laboratory experiments.

Nervous System Processes
Everyday situations can highlight the complex action of the nervous system. For example, in a soccer match, players anticipate the opportunity to act. At the exact moment the ball is thrown in a player's direction, thousands of nerve connections start to become active. In milliseconds, the player begins to use sensory memories and visual information to immediately decide the best course of action, such as to kick the ball to another player or directly to the goal. The central nervous system consists of the brain and spinal cord. The brain is the control central of the nervous system. The spinal cord conducts electrical signals between the brain and various nerves throughout the body, and controls some reflex functions. Neurons are cells that propagate the electrical impulses in the nervous system, and glial cells help maintain parts of the nervous system. For example, they produce myelin, which coats many neurons like insulation in electrical wiring. The neurons have important properties, such as excitability and conductivity, and act similarly to an electric current transmitted along a wire. This phenomenon occurs because of permeation of ions, such as sodium and potassium, through the neural membrane, which generate an electrical signal that propagates between neurons via its branched structure, consisting of thousands of small extensions.

Early Research

Nerve impulse propagation and the nervous system processes have been researched for many years using theories and techniques from genetics, molecular biology, physiology, psychology, and mathematics, among others. In the 1950s, physiologists and biophysicists Alan Hodgkin and Andrew Huxley experimented on the nervous systems of squids, specifically on a structure known as the "giant axon." An axon transmits electrical impulses in the nervous system, and a squid's giant axon can be up to 1 millimeter in diameter, much larger than most axons and visible to the naked eye. These experiments led to the development of the Hodgkin–Huxley equations, which are nonlinear ordinary differential equations that describe or approximate the electrical characteristics of neurons and other electrically excitable cells, such as those in the heart. They involve concepts like gates (channels that allow the ions to flow), voltage thresholds, and conductances, which act together to determine if and when a neuron "fires" an electrical burst. They are very similar to electric circuit theory, and some models of nervous systems look very much like electrical circuit diagrams.

Other Mathematical Connections

Hodgkin and Huxley won a Nobel Prize for their experimental and mathematical work, which has since led to other mathematical explorations of the nervous system. The nervous system in mammals is a very complex dynamic system, with many interconnected components. Periodic rhythms are found in some types of movement-related behaviors that are governed by the nervous system, like walking and breathing. They are also related to sensation and cognition. Studies of all these various substructures involve not only understanding how each structure behaves on its own but also how they interconnect and communicate with one another. Because of the vast degree of intercorrelation among various nervous system structures, from individual neurons to larger structures like the brain and spinal cord, one challenge facing mathematical modelers is creating systems of equations that optimize the ability of the equations to realistically represent neuronal systems and their behaviors while making them tractable for computation and interpretation. One of the interesting mathematical phenomena that researchers study is called "gamma and beta rhythms." These brain waves have been connected to so-called "higher" mental activity, like perception and consciousness, as well as to synchronous activity that may help link various sensory inputs into a single mental construction of an object. However, many questions remain. Techniques such as graphs, circuits, networks, clustering, geometry, and simulation all play a role in investigation of nervous system properties and functions.

One additional important advance in neuroscience is the neurochip. It can be used to help link biological neurons and semiconductor materials, which may one day help to create prosthetics that integrate fully into the body's own neural system. They may also facilitate treatments for neurological diseases like Alzheimer's and Parkinson's.

Further Reading

Ermentrout, G. B., and D. H. Terman. *Mathematical Foundations of Neuroscience (Interdisciplinary Applied Mathematics)*. New York: Springer, 2010.

Gabbiani, F., and S. T. Cox. *Mathematics for Neuroscientists*. Oxford, England: Elsevier, 2010.

Kopell, Nancy. "We Got Rhythm: Dynamical Systems of the Nervous System." *Notices of the American Mathematical Society* 47, no. 1 (January 2000). http://www.ams.org/notices/200001/fea-kopell.pdf.

Scott, Alwyn. *Neuroscience: A Mathematical Primer*. New York: Springer, 2002.

Maria Elizete Kunkel
Maria Elizabeth S. Rodrigues

Nutrition

Category: Medicine and Health.
Fields of Study: Algebra; Data Analysis and Probability; Number and Operations.
Summary: Mathematicians and nutrition scientists model and analyze numerous aspects of nutrition and diet.

Nutrition is the system of providing food to an organism as well as the science of food and eating. Nutrition science is an interdisciplinary field that involves a wide variety of disciplines, including mathematics, statistics, culinary science, physiology, genetics, biochemistry, psychology, medical sciences, sociology, anthropology, and ethnography. Mathematical and statistical methods are widely used to describe and analyze different nutrients in food, determine their impact on nutrition and health, develop eating plans, assess public opinion, and inform public policy. Meal-planning and nutritional labels are often used in classroom mathematical problems.

Taxonomies of Human Nutrients

The understanding of what balanced nutrition means is a difficult and controversial subject, with many schools of scientific thought, cultural traditions, and governing bodies proposing different ways of eating. Many people may be familiar with the U.S. Department of Agriculture's food pyramid, which was introduced in the 1990s to replace the older four food groups model. There are similar guidelines produced in many countries, using a mix of scientific research and expert opinion. One measurement scale for a nutrient is the time it takes for its lack to manifest itself in health problems. Lack of energy-providing nutrients, such as fats, proteins, and carbohydrates, is felt within hours as hunger and causes symptoms within days. A disbalance of macrominerals, such as potassium and calcium, as well as some vitamins has been shown to lead to specific diseases and is felt within weeks to months.

For example, physician James Lind published an influential treatise on scurvy in 1753, based in part on his controlled experiments with British sailors. Lack of some other vitamins and minerals can take years to manifest. Flavonoids are found in plants, changing coloration, smell, and taste. Researchers hypothesize that flavonoids regulate organism responses, such as inflammations and allergies, as well as reactions to carcinogens, bacteria, and viruses. Most studies of flavonoids in human nutrition are only decades old. Probiotics are live microorganisms most frequently eaten with fermented food. They may affect the immune system, blood pressure, inflammation, and cancer. Prebiotics are food items, such as inulin in chicory roots, that promote the growth of microorganisms in the digestive tract. Water participates in most systems and processes in the body, playing roles such as a solvent for other substances and entering chemical reactions.

Researching and Modeling Nutrition

Data collection and quantitative analysis are used for a variety of purposes in nutrition science. For example, they are used to investigate the effects of nutritional deficiencies, optimize diets for long-term health and longevity, study the effectiveness of weight-loss plans, and establish causal links between political policy changes and nutrition and health effects. Qualitative methods like case studies and ethnographic studies are insufficient to establish cause, though they may highlight key variables. One critical principle of scientific studies that seek to make causal connections is isolating a small number of variables to systematically manipulate, while controlling the rest. Because nutrients interact with all systems in the body—with other organisms living in the body, with each other, and with behaviors other than eating—the complexity of the resulting system can make this approach difficult. Further, the effects of some types of nutrients take years or even decades to uncover, or they may occur

The Japanese traditionally eat more fish, vegetables, grains, and fruit and consume smaller portions than most Western diets. (Photos.com)

in only a small number of people. Studying these would require extensive longitudinal studies or very large sample sizes to be statistically valid, which may have significant practical and ethical barriers. Finally, individual differences in reactions to nutrition changes may be large and non-random, depending on genetics, culture, and daily habits, which means that averaging the effects of nutritional interventions may overlook important effects on small minorities, such as allergic reactions.

Mathematicians and nutrition scientists use mathematical modeling and simulation to investigate the functions of systems and to experiment with the conditional responses of multiple variables. Increases in computing power have made complex modeling a feasible alternative to traditional scientific experimentation. Problems are drawn from areas of concern, such as obesity, diabetes, cancer, and toxicology. Many models rely on collection of kinetic body data to develop accurate models of physiological processes, such as bioperiodicity and membrane transport, which is also possible because of advances in medical imaging and other technology. Computational approaches are used to estimate distributions of parameters, evaluate linear integrators and other functions, manipulate multiple variables in stochastic models, and create visualizations. Mathematical or statistical approaches, such as neural networks, graph theory, and cluster analysis, have also been used to model data or systems and to make connections.

Genetically modified foods are a controversial subject in nutrition. Typical reasons for altering food are for resistance to pests or disease or for nutritional benefits. The Swiss-developed "golden rice" has higher levels of vitamin A than standard rice strains, which would theoretically benefit third world countries where rice is a staple food and vitamin A deficiencies are common. Some support the use of such foods to combat hunger in areas of the world with chronic shortages and endemic malnutrition. Others cite the unknown long-term effects, such as spontaneous cross-pollination with unmodified organisms, as well as the ethical implications. Mathematicians and scientists have helped to create genetically modified foods and have investigated many questions related to them. For example, informaticists have used combinatorial reduction rules to create a model to detect unknown, genetically modified organisms. Others research and model aspects such as the likelihoods of positive and negative ecological outcomes, pathogenicity, public acceptance, and impacts on international trade using probabilistic and statistical methods, simulation, differential equations, and a wide variety of computerized modeling techniques.

Diets and Meal Planning

A diet is the description of types and quantities of nutrients consumed. Because organisms vary in ways other than food intake, dietary variables are typically studied in their relationships with other variables—either direct proportionality or more complex functions. Different cultures have varied proportions of nutrients in their diets, as well as certain prohibitions. For example, Aleuts traditionally eat a large amount of meat, consuming about eight times more protein than South American agricultural tribes. Japanese and Mediterranean diets are often cited for emphasis on certain fats, fruits, vegetables, and carbohydrates. Both Jewish and Muslim traditions forbid certain types of foods. People may also choose diets for specific goals, such as weight loss or control of medical conditions like diabetes or high blood pressure, often with little scientific evidence of effectiveness—though scientists are seeking ways to validate or refute such claims. Globalization has made different types of diets and foods increasingly known and accessible to people everywhere.

Software for planning least-cost nutritional meals was developed for mainframe computers in the early 1960s and evolved during the 1970s to include food preferences options. Later research in the 1980s and the evolution of personal computing led to new software that used mathematical programming to optimize and maximize menu planning for different variables, including nutrition, allergies, and preferences. Internet-based software and algorithms, such as that used by the weight-loss company Weight Watchers with their Weight Watchers Online program, now allows people to track and plan menus based on a variety of criteria, often dynamically linked to databases with recipes, past behavior, and weight or measurement tracking. Large institutions, such as schools and hospitals, may use software that includes inventory and other supply variables.

Nutrition and Mathematical Problem Solving

There are studies directly linking nutrition and success in mathematics. One group of researchers found that providing a balanced breakfast before the morning mathematics class raised test scores more than any other variable they analyzed, such as changes in teaching methods. Different cultures have different beliefs of what constitutes "brain food." Certain types of fat, vitamins B and C, and monosaccharides have been shown to increase memory and speed of computation within time periods from minutes to days from increased consumption. More complex cognitive effects of food, such as connections between gluten or lactose sensitivity and attention, are being investigated.

Further Reading

Bhargava, A. *Econometrics, Statistics and Computational Approaches in Food and Health*. Singapore: World Scientific Publishing, 2006.

Kowtaluk, Helen. *Discovering Food and Nutrition*. New York: McGraw-Hill, 2004.

Noss Whitney, Eleanor, and Sharon Rady Rolfes. *Understanding Nutrition*. Belmont, CA: Wadsworth Publishing, 2010.

Novotny, Janet, Michael Green, and Ray Boston. *Mathematical Modeling in Nutrition and the Health Sciences*. New York: Springer, 2003.

Maria Droujkova

Pacemakers

Category: Medicine and Health.
Fields of Study: Algebra; Geometry.
Summary: Artificial pacemakers send a signal to the heart to keep it pumping and mathematicians develop models to determine when and how often to do so.

While a pacemaker is often thought of as a regulator for the heart, a variety of natural pacemakers are responsible for regulating numerous bodily functions including circadian rhythms and menstruation. The actions of natural pacemakers can be modeled as coupled oscillators, where, for example, the behavior of the natural pacemaker influences the function of the heart and vice versa. Square waves or sine waves are often useful in understanding the theory of coupled oscillators, which dates back to 1665 when Christiaan Huygens noticed synchronization in pendulum clocks.

Scientists and mathematicians have shown that chaotic oscillation or amplitude death can also occur in coupled scenarios. A change in the rhythm or in the way they are coupled can result in a change in function, such as in irregular menstrual periods or menopause. Dynamical systems model the interactions between coupled oscillators and allow for theoretical predictions. Using these models, mathematicians, biologists, and medical professionals have made significant advances in understanding natural pacemakers and in designing effective artificial pacemakers. Some of the related mathematical theory is taught to undergraduate mathematics students.

Heart Rhythms and Pacemakers

The sinoatrial node (SA node) is thought to act as the heart's natural pacemaker via electrical impulses. The typical rate for a resting heart is 60 to 70 beats per minute. The pacemaker cells keep the heart pumping at a steady rate, but medical problems can lead to chaotic behavior and cardiac arrest.

Defibrillation may reset the rhythm in some cases but an artificial cardiac pacemaker may be required if the rhythm remains chaotic. Wavelet transforms have been used to effectively model cardiac signals but implementation is difficult because of high power consumption. Australian anesthesiologist Mark Lidwell and physicist Edgar Booth are believed to have designed the first artificial pacemaker in 1928.

American physiologist Albert Hyman also developed an early pacemaker. Many designers of artificial pacemakers have assumed that regular impulses from a pacemaker should be used to stabilize the heartbeat. However, a periodic signal may lead to chaos in some mathematical models, so scientists are developing pacemakers that send impulses based on chaos control theory.

Body Clock and Jet Lag

Jet lag is thought to to result from a desynchronization of the suprachiasmatic nucleus (SCN) pacemaker cells in the hypothalamus of mammals. Experimental studies suggest that the SCN may synchronize within one week. Scientists and mathematicians have mathematically modeled the system as a network with connections between the cells, which are called *nodes* in the language of graph theory.

For example, mathematicians Channa Navaratna and Menaka Navaratna have adapted a model of neuroscientist Peter Achermann and bioinformaticist

Hanspeter Kunz. The hypothalamus is thought to have 16,000 pacemaker cells, so they analyzed computer data from a model with this many pacemaker cells and found that the number of long-distance connections in the network determined the synchronicity time. They examined the types of network connections that are needed between the nodes in order to make the model synchronize in a week, and they designed a model that consistently synchronized in close to seven days.

Scientists and mathematicians have also studied many other issues related to pacemakers, such as interference and power issues. There is controversy and conflicting evidence on whether devices such as cell phones or iPods affect pacemakers. Many medical professionals presume an association until clearer evidence to the contrary is found and recommend keeping the devices at least a few inches away from a pacemaker to err on the side of caution. Scientists have developed what some call "origami batteries" made of carbon nanotubes and cellulose that may power the next generation of pacemakers. The batteries can be cut into many shapes.

Further Reading

Barold, S. Serge, et al. *Cardiac Pacemakers Step by Step: An Illustrated Guide.* Hoboken, NJ: Wiley, 2003.

Glantz, Stanton. *Mathematics for Biomedical Applications.* Berkeley: University of California Press, 1979.

Strogatz, Steven. *Nonlinear Dynamics and Chaos: With Applications to Physics, Biology, Chemistry, and Engineering.* Boulder, CO: Westview Press, 2001.

Sarah J. Greenwald
Jill E. Thomley

Planetary Orbits

Category: Space, Time, and Distance.
Fields of Study: Algebra; Geometry.
Summary: It took mathematicians thousands of years to accurately describe planetary motion.

For millennia, the shape of the paths in which the planets orbited was dominated by metaphysical concerns and assumed, almost without question, to be circular. It was not until the seventeenth century that science discovered the actual shape of planetary orbits, the ellipse.

Early Conceptions

In ancient Greek astronomy, it was assumed that the Earth was the center of the universe, and all of the known planets (including the sun and the moon) as well as the stars revolved around it. Furthermore, at least from the time of Pythagoras (c. 569–475 b.c.e.), these orbits were assumed to be circular. This assumption was a metaphysical one.

The Pythagoreans believed in the perfection of mathematics and held the view that the circle was perfect because of its symmetry and continuity. Therefore, the universe must surely be constructed to reflect this perfection by requiring the planets to revolve around the Earth in perfect circular motion. That influential philosophers such as Plato and Aristotle accepted the perfection of circular motion contributed to the fact that the idea went almost unchallenged for nearly 2000 years.

With the increasing ability to make accurate observations of the movements of the heavens and mathematical calculations to predict those movements, the simple assumption of perfect circular motion became more problematic. The predictions of the planetary positions did not match the actual observed locations. Eudoxus (408–355 b.c.e.) addressed this discrepancy by devising a complicated system of nested spheres in which each planet moved, maintaining circular motion of each sphere while more accurately predicting the location of the planets.

For many centuries, one man's work dominated European thinking on planetary motions. The Greek mathematician and astronomer Ptolemy (85–165 c.e.) compiled all that was known about the movements of heavenly bodies into one work that came to be known as *The Almagest*. This book employed an array of very complex geometric and trigonometric theories to describe the movement of the planets, with the Earth remaining at the center. In order for the observations to be as close as possible to the calculations, Ptolemy used epicycles (small circles revolving upon bigger circles as they revolve around the Earth) and moved the Earth away from the center of revolution of the planets.

The new center of revolution was an imaginary point some distance away from the Earth. Ptolemy's influence on Western astronomy was partially because of its general agreement with Christian doctrine. As the center of God's creation, the Earth must rest at the center of the cosmos. Furthermore, a perfect Creator would use the perfect circle to put His creation in motion.

Challenges

The most serious challenge to Ptolemaic cosmology came from the Polish church official, Nicolaus Copernicus (1473–1543), whose revolutionary work *De Revolutionibus* placed the sun, not the Earth, at the center of the universe, relegating the Earth to mere planethood. Copernicus, however, remained adamant in his belief that the planets orbited the sun in a composite of perfect circular motions. The doctrine of perfect circular motion in the heavens was finally challenged by the German astronomer Johannes Kepler (1571–1630). Kepler, after many years of tedious and painstaking calculations involving the orbit of Mars, finally determined that Mars actually orbited the sun in an elliptical orbit, not a circular one. This revolutionary idea was based in part on another discovery by Kepler that the speed of the planets varied as they orbited the sun. Later, the great British mathematician and scientist, Isaac Newton (1643–1727), used his universal law of gravitation and laws of motion to provide a mathematical explanation for Kepler's claim of elliptical orbits, finally putting an end to the ancient doctrine of circular motion in the heavens.

Mathematics continues to play an important role in modeling planetary orbits. For example, Mercury's orbit is more accurately represented with hyperbolic geometry than with Euclidian geometry. Further, the orbit of Mercury allows researchers to see the impact of the sun's gravitational field on the curvature of space.

Further Reading

Danielson, Dennis Richard. *The Book of the Cosmos: Imagining the Universe From Heraclitus to Hawking*. Cambridge, MA: Perseus Publications, 2000.

Gingerich, O. *The Eye of Heaven: Ptolemy, Copernicus, Kepler*. New York: American Institute of Physics, 1993.

Heath, Thomas L. *Greek Astronomy*. New York: E. P. Dutton, 1932.

Kopache, Gerald. "Planetary Motion: Also Featuring Some Stars, Some Comets and the Moon." Ceshore Publishing Company, 2004.

Montenbruck, Oliver, and Gill Eberhard. *Satellite Orbits: Models, Methods and Applications*. Berlin: Springer, 2000.

Pannekoek, Anton. *A History of Astronomy*. New York: Dover Publications, 1989.

Sagan, Carl. *Cosmos*. New York: Random House, 1980.

Todd Timmons

Plate Tectonics

Category: Weather, Nature, and Environment.
Fields of Study: Data Analysis and Probability.
Summary: Tectonic plate movement is measured and analyzed using mathematics.

The ideas of plate tectonics and continental drift have been theorized by many scientists over the years. For example, in the early twentieth century, Alfred Wegener publicly presented theories regarding the existence of a supercontinent called "Pangea" that eventually formed all the known continents. He supported centrifugal force as an explanation for drift. A few years later, Arthur Holmes supported thermal convection as an explanation. At the time, there was insufficient mathematical and scientific evidence to support these theories and they were largely dismissed, in part because seeing into the depths of the oceans and into the Earth itself is often a more difficult venture than seeing galaxies at the far reaches of the universe. By the latter half of the twentieth century, discoveries such as mid-Atlantic underwater volcanic chains and the mapping and mathematical analysis of seismic activity suggested the existence of large, mobile plates in the Earth's crust.

In the twenty-first century, scientists and mathematicians are still developing new and innovative ways to collect data, model, visualize, and simulate the Earth's inner structure. For example, geophysicist Robert van der Hilst and mathematician Maarten Van de Hoop have used a mathematical technique known as "microlocal analysis," as well as statistical methods, such as confidence intervals, to explore the geometry of the layers near the boundary of the Earth's core and mantle. This technique extends existing methods for analyzing noisy seismic data. It produces not only an image, but also an estimate of the probability that a true layer has been discovered. Ongoing collaboration between mathematicians and geophysical scientists is crucial to address the massively scaled problems that arise in geoscience, such as continental drift. This is true not only for data collection in the field, but also for computer simulation, which is increasingly an avenue of exploration and cross-validation for theories and data. These simulations often require combining many scales of data, both macro and micro, as well as observations collected over different periods of time. Further, much of the data is noisy, incomplete, or difficult to directly measure. Mathematics is also involved in the increasingly sophisticated tools that allow scientists to visit the depths of the oceans and begin to look at some previously impenetrable layers of the Earth.

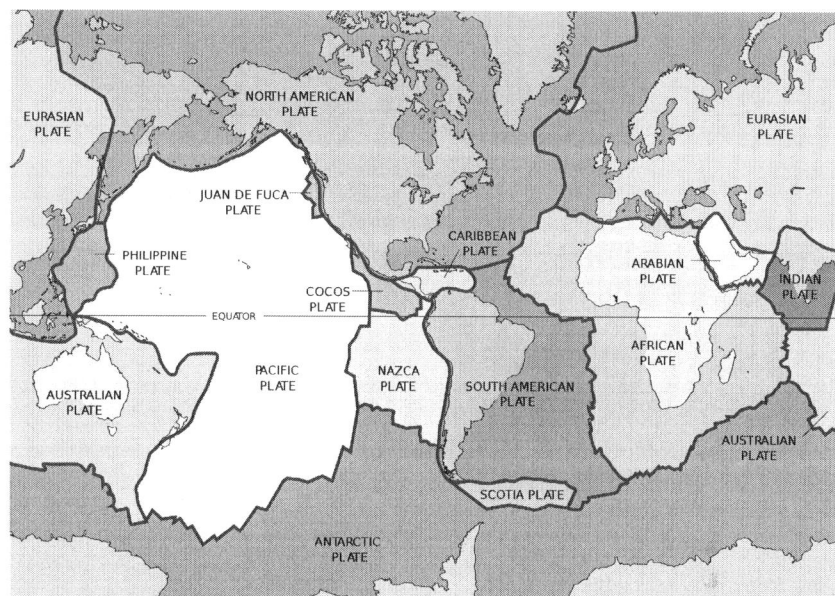

A U.S. Geological Survey illustration of Earth's rigid slabs (called tectonic plates) that are moving relative to one another. (U.S. Geological Services)

The Spreading Sea Floor

As an officer in the U.S. Navy, Harry Hess's curiosity led him to measure the ocean floor using sounding gear and magnetometers during World War II. Once the war ended, Hess developed the theory of sea floor spreading to explain his data. He proposed that magma oozed up between the plates along the ridges in the ocean floor, pushing them apart and causing the plates to move.

Strips of rock parallel to the ridges provide evidence for sea floor spreading. Strips closest to the ridge have the same polarity as the Earth (magnetic north pointing to the north pole); however, the strips moving out away from the ridge on opposite sides mirror each other and alternate between current polarity and reversed polarity as the Earth's magnetic field reversed over time. These alternating strips suggest that new rock is created along the ridges over geologic time.

Continents Adrift

Until 1912, scientists assumed that the continents were fixed in place. In that year, Alfred Wenger suggested that the continents were adrift, originally part of one large landmass. Wegner cited evidence such as matching geological formations and fossils from South America and Africa. It was not until the late 1960s that discoveries were made and measuring techniques improved to the extent that the theory of plate tectonics emerged and became widely accepted. Scientists now recognize that the continents are attached to plates and move with them rather than moving independently. Scientists also now know that the plates that make up Earth's crust and the continents attached to them are moving several centimeters per year on average as they collide, move apart, and brush up against each other.

Plate Movement

Muawia Barazangi and James Dorman (1969) charted the locations of all earthquakes occurring from 1961 to 1967 and found that most occurred in a narrow band of seismic activity. This band of high earthquake and volcanic activity, commonly called the "Pacific Ring of Fire," defines many plate boundaries around the Pacific Ocean.

Most plate movement occurs along the edges of the plates. Scientists can measure the velocity (speed and direction) of plate movement and determine how that relates to earthquake and volcanic activity. For historical information, scientists turn to ocean floor magnetic striping data and geological dating of rock formations.

Measurement techniques have improved greatly since Hess's measurements. The most common technique for measuring plate movement in the early twenty-first century is the Global Positioning System (GPS). As satellites continuously transmit radio signals to Earth, each GPS ground site simultaneously receives signals from at least four satellites. By recording the exact time and location of each satellite when its signal was received, it is possible to determine the precise position of the GPS ground site on Earth (longitude, latitude, and elevation). Regularly measuring distances between specific points allows scientists to determine if there has been active movement between plates on a scale of millimeters. Using time-series graphs and plotting vectors, it is possible to learn how the plates move.

While scientists know that most earthquakes and volcanoes occur along plate boundaries, they still cannot predict exactly when and where they will occur. By monitoring plate movement, scientists hope to learn more about the events building up to earthquakes and volcanic eruptions.

Further Reading

Barazangi, Muawia, and James Dorman. "World Seismicity Maps Compiled from ESSA, Coast and Geodetic Survey, Epicenter Data, 1961–1967." *Bulletin of the Seismological Society of Am*erica 59 (1969).

Preskes, Naomi. *Plate Tectonics: An Insider's History of the Modern Theory of the Earth.* Boulder, CO: Westview Press, 2003.

Christine Klein

Predator–Prey Models

Category: Weather, Nature, and Environment.
Fields of Study: Algebra; Data Analysis and Probability; Number and Operations.
Summary: The interaction between the population sizes of a predator species and a prey species can be modeled using systems of equations.

Predator–prey models are systems of mathematical equations that are used to predict the populations of interacting species, one of which—the prey—is the primary food source for the other—the predator. One famous example that has been extensively studied is the relationship between the wolves and moose on Isle Royale in Lake Superior.

The Isle Royale populations are well suited for modeling the predator–prey relationship because there is little food for the wolves other than the moose and there are no other predators for the moose. In addition, the geographic isolation limits other factors that would complicate the mathematics in the equations, such as hunting or migration. This predator–prey interaction has been carefully studied since the 1950s and continues to be investigated into the twenty-first century.

Modeling Predator–Prey Populations

Most predator–prey models are composed of two equations, the first representing the change in the prey population, and the second the change in the predator population. Each equation has the following form: birth function minus death function.

If $X(t)$ represents the quantity of prey at time t, and $Y(t)$ represents the quantity of predators at time t, then the instantaneous rate of change in prey is

$$\frac{dX}{dt} = f_1 - f_2$$

and the instantaneous rate of change in predators is

$$\frac{dY}{dt} = f_3 - f_4$$

where f_1 is the mathematical term that describes the births in the prey population, f_2 describes the deaths in the prey population, f_3 describes the births in the predator population, and f_4 describes the deaths in the predator population.

There have been many predator–prey models proposed since the beginning of the twentieth century. The most famous and the earliest known is the Lotka–Volterra system, named for the two scientists who developed the same mathematical model independently, American Alfred Lotka (1880–1949) publishing the equations in 1925 and Italian Vito Volterra (1860–1940) publishing them in 1926. Lotka had degrees in physics and chemistry, and he believed that one could apply physical principles to biological systems. His work on predator–prey interactions is just part of extensive work he published in 1925 in the text titled *Elements of Physical Biology*. Lotka used a chemical reaction analogy to justify the terms in the model.

In the absence of predators, the prey should increase at a rate proportional to the current quantity of prey, X. In other words, more moose around to mate without being hunted means more calves would be born. Likewise, in the absence of prey, the predators should die off at a rate proportional to the current predator population, Y. In other words, with many wolves and no moose for food, more wolves would starve.

Lotka used a chemical reaction analogy to explain prey deaths and predator births: when a reaction occurs by mixing chemicals, the rate of the reaction is proportional to the product of the quantities of the reactants. Lotka argued that prey should decrease and predators should increase at rates proportional to the product of the quantity of prey and predators, XY. In other words, the moose deaths should be closely related to the rate of interaction of wolves and moose, and the wolf births should be as well because wolves need the moose for food to be healthy and have pups. The equations can be written as

$$\frac{dX}{dt} = aX - bXY \text{ and } \frac{dY}{dt} = cXY - dY$$

for non-negative proportionality constants a, b, c, and d.

Volterra arrived at the same model using different reasoning. Volterra was a physicist whose daughter and son-in-law were biologists. While looking for a mathematical explanation for a problem his son-in-law was working on, Volterra became very interested in interactions of species and spent the rest of his professional life looking for a mathematical theory of evolution.

The Lotka–Volterra predator–prey model can be solved without a computer and yields a graph that makes sense. The population of the predator oscillates as does that of the prey, with the predator population trailing slightly behind. Too many prey results in more predators, who swamp the prey causing a decrease in prey. As the prey become scarce, the predators also start to die out, and the cycle begins again (see Figure 1).

While this result has reasonable qualitative behavior, many scientists have objected to the equations in this form. Some of the concerns about the model have included the following:

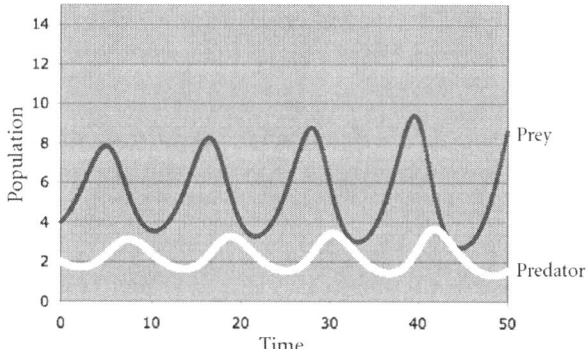

Figure 1. Predator–prey interaction.

- If there are no predators, the prey population would grow arbitrarily large
- A reduction in the number of prey should cause more predator deaths rather than fewer predator births
- For a fixed number of predators, the number of prey eaten is proportional to the number of prey present, implying that predators are always hungry and eat the same proportion of the prey no matter how large the number of prey gets
- The food for the prey plays a role in the births and deaths of the prey, and should be included in the model
- No spatial considerations are incorporated in the model, so factors such as migration or seeking safety in herds are ignored
- These equations do not take into account gestation periods and seasonal changes in birth rates
- The constants a, b, c, and d are difficult to estimate for a given situation without a large amount of data collected from field observations

Much work has been done since the 1930s to modify the equations to address these concerns and to apply the equations to data from specific situations, such as the moose and wolves of Isle Royale. In the twenty-first century, scientists use sophisticated computer models to model predator–prey interactions using increasingly intricate equations to incorporate more realistic assumptions in the mathematics.

Further Reading

Kingsland, Sharon E. *Modeling Nature, Episodes in the History of Population Ecology*. Chicago: University of Chicago Press, 1995.

Lotka, Alfred J. *Elements of Physical Biology*. Baltimore, MD: Williams and Wilkins Publishers, 1925.

Volterra, Vito. "Variations and Fluctuations of the Number of Individuals in Animal Species Living Together." In *Animal Ecology*. Edited by R. Chapman. New York: McGraw-Hill, 1926.

Vucetich, John A. "The Wolves and Moose of Isle Royale." http://www.isleroyalewolf.org/wolfhome/home.html.

Holly Hirst

Pregnancy

Category: Medicine and Health.
Fields of Study: Algebra; Data Analysis and Probability.
Summary: Various mathematical models help describe issues related to conception, diseases associated with pregnancy, and population dynamics.

Much of the conclusions drawn in medicine, in particular in obstetrics and gynecology, are often based on heuristics, limited observations, and sometimes even biased data. Mathematicians and statisticians have recently attempted to develop general theoretical models that can be adapted to specific situations in order to facilitate the understanding of various aspects of human pregnancy. Specifically, more recent studies have been conducted regarding conception time, disease prediction related to pregnancy, and the effect of pregnancy on population growth.

Modeling the Most Efficient Time to Conceive

One of the most fundamental and important research topics in the study of human pregnancy is the so-called time-to-pregnancy (TTP). TTP can be defined scientifically as the number of menstrual cycles it takes a couple engaging in regular sexual intercourse with no contraception usage to conceive a child. Fittingly, statisticians attempt to generate as much data as possible from various couples regarding their personal TTP experiences. The data are collected in a way that is as unbiased as possible—it is intended to accurately represent couples in the general population attempting to conceive a child. From the data, both qualitative and quantitative statistical methods are implemented in order to ascertain the most efficient method to achieve conception.

For example, some social trends increase the age at which a woman attempts to become pregnant. When this situation arises, women are often concerned about achieving conception before the onset of infertility, which proceeds menopause. In fact, couples that are unsuccessful in conceiving within one year are clinically classified as *infertile*. When this condition occurs, medical doctors often recommend that the couple engage in assisted reproductive therapy (ART). However, ART can be very expensive and often increases the risk of adverse outcomes for the offspring, including various birth defects. Therefore, statistical models have been developed that pose an alternative to ART. These models are developed using Bayesian decision theory, named for Thomas Bayes, and search for optimal approaches for a couple to time intercourse in order to achieve conception naturally, without the potentially disadvantageous ART. These models quantitatively incorporate various biological aspects, including menstrual cycles and basal body temperature, as well as the monitoring of electrolytes—among other phenomena—in order to be as efficient as possible.

Predicting Diseases Associated With Pregnancy

Medical evidence supports the notion that women often repeat reproductive outcomes. In particular, women with a history of bearing children with adverse outcomes often have up to a two-fold increase in subsequent risk. Therefore, researchers in the mathematical and statistical sciences realized the necessity for statistical analyses that address this issue. In fact, statistical research has been conducted in order to promote a consistent strategy that assesses the risks each woman may face in a subsequent pregnancy. The goal is for these types of models to become increasingly more accurate, as they incorporate statistical data regarding the recent reproductive history of the woman, among other biological factors, which were not fully taken into account in previous studies.

Mathematical epidemiology (the study of the incidence, distribution, and control of diseases in a population) attempts to better comprehend, diagnose, and predict various diseases incorporated with pregnancy, and this field is ever-expanding. By designing and implementing various statistical approaches and mathematical models to better predict realistic outcomes, mathematicians and statisticians have studied congenital defects and growth restrictions, as well as preterm delivery, pre-eclampsia, and eclampsia.

For example, pre-eclampsia is a pregnancy condition in which high blood pressure and high levels of protein in urine develop toward the end of the second trimester or in the third trimester of pregnancy. The symptoms of this condition may include excessive weight gain, swelling, headaches, and vision loss. In some cases this condition can be fatal to the expectant mother or the child. The exact causes of pre-eclampsia are unknown at the beginning of the twenty-first century, and the only cure for the disease is the delivery of the child. Therefore, it is apparent that determining which women are prone to develop pre-eclampsia is an exceedingly important area of research.

Empirical evidence indicates that a woman's heart rate is a deterministic factor in the prediction of pre-eclampsia. In recent times, statisticians have therefore developed a novel and non-invasive approach to detect abnormalities in pre-eclamptic women that distinguishes from women with non-pre-eclamptic pregnancies. This approach is accomplished by comparing the dynamical complexity of the heart rates of women that are pre-eclamptic with those that are non-pre-eclamptic. The analysis revealed that the heart rate of pre-eclamptic women demonstrated a more regular dynamic behavior than those women that were not pre-eclamptic, which substantiates the empirical notion that diseased states may be associated with regular heart rate patterns.

Population Dynamics

Mathematicians have long developed models to analyze population dynamics. One contemporary model also incorporates how pregnant women directly influence such dynamics. This model consists of an equation that describes the evolution of the entire population and an equation that analyzes the evolution of pregnant women. These equations are coupled—they are studied simultaneously. Moreover, this particular system of equations can be analyzed as a linear model (not sensitive to initial data), with or without diffusion

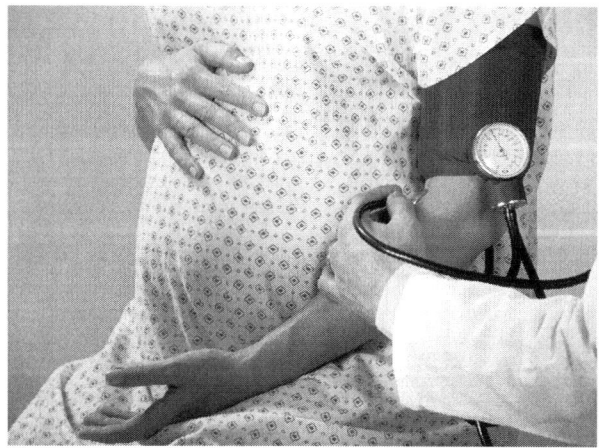

Mathematical epidemiology is used to predict conditions like pre-eclampsia during pregnancy. (Photos.com)

(permitting members of the population to travel large distances), or as a nonlinear model (sensitive to initial data) without diffusion. The asymptotic behavior of the solutions to this system (the long-term behavior of the population) was also addressed.

Further Reading

Fragnelli, Ginni, et al. "Qualitative Properties of a Population Dynamics System Describing Pregnancy." *Mathematical Models and Methods in Applied Sciences* 4 (2005).

Germaine B., et al. "Analysis of Repeated Pregnancy Outcomes." *Statistical Methods in Medical Research* 15 (2006).

Salazar, Carlos, et al. "Non-Linear Analysis of Maternal Heart Rate Patterns and Pre-Eclamptic Pregnancies." *Journal of Theoretical Medicine* 5 (2003).

Savitz, David A., et al. "Methodologic Issues in the Design and Analysis of Epidemiologic Studies of Pregnancy Outcome." *Statistical Methods in Medical Research* 15 (2006).

Scarpa, Bruno, and David B. Dunson. "Beyesian Methods for Searching for Optimal Rules for Timing Intercourse to Achieve Pregnancy." *Statistics in Medicine* 26 (2007).

Scheikle, Thomas H., and Niels Keiding. "Design and Analysis of Time-to-Pregnancy." *Statistical Methods in Medical Research* 15 (2006).

Daniel J. Galiffa

Psychological Testing

Category: Medicine and Health.
Fields of Study: Algebra; Data Analysis and Probability; Representations.
Summary: Though they often require a subjective element, psychological tests make every effort to generate useful quantitative data.

Testing is used for many different purposes within psychology—among them to evaluate intelligence, diagnose psychiatric illness, and identify aptitudes and interests. Although the results of testing are rarely used as the sole criterion to make a diagnosis or other decision about an individual, they are often used in conjunction with information gained from other sources, such as interviews and observations of behavior. There are many types of psychological tests, but most share the goal of expressing an essentially unobservable quality such as intelligence or anxiety in terms of numbers. The numbers themselves are not meant to be taken literally—no one seriously believes that a person's intelligence is equivalent to their IQ score, for instance. Instead the numbers are useful tools that help evaluate a person's situation; for instance, how does the intellectual development of one particular child relate to that of other children of his age? Of course, the results of psychological testing should be evaluated with the social context of the individual in mind and with full respect for human diversity.

Psychometrics

Psychometrics is a field of study that applies mathematical and statistical principles to devise new psychological tests and evaluate the properties of current tests. Psychologist Anne Anastasi was often known as the "test guru" for her pioneering work in psychometrics. In her 1954 book *Psychological Testing*, she discussed the ways in which trait development is influenced by education and heredity as well as how differences in training, culture, and language affect measurement. The two most common approaches to psychometrics in the twenty-first century are classical test theory and item response theory (IRT).

Classical test theory is the older approach and the calculations required can be performed with a pencil and paper, although twenty-first-century computer software is often used. Classical test theory assumes that all measurements are imperfect and thus contain error: the goal is to evaluate the amount of error in a measurement and develop ways to minimize it. Any observed measurement (for instance, a child's score on an intelligence test) is made up of two components: true score and error. This may be written as an equation: $X = T + E$, where X is the observed score, T is the true score (the score representing the child's true intelligence), and E is the error component (resulting from imperfect testing). Classical test theory assumes that that error is random and thus will sometimes be positive (resulting in a higher observed score than true score) and sometimes negative (resulting in a lower observed score than true score) so that over an infinite number of testing occasions, the mean of the observed scores will equal the true score. Although normally a test is administered only once to a given individual, this is a useful model that facilitates evaluation of the reliability and validity of different tests.

Item response theory (IRT) is a different approach to psychological testing and assumes that observed performance on any given test item can be explained by a latent (unobservable) trait or ability so that individuals may be evaluated in terms of the amount of that trait they contain, and items may be evaluated in terms of the amount of the trait required to answer them positively. For an item on an intelligence test (intelligence being the latent trait), persons with higher intelligence should be more likely to answer the question correctly. The same principle applies to IRT-based tests evaluating other psychological characteristics; for instance, if an item in a psychological screening test is meant to diagnose depression, a person with more depressive symptoms should be more likely to answer it positively. IRT is a mathematically complex method of analysis that depends on the use of specialized computer software and has become a popular means to evaluate psychological tests as computers have become more affordable. Although the mathematical models of IRT differ from that of classical test theory, the goals are the same: to devise tests that measure characteristics of individuals with a minimum of error.

Reliability and Validity

The term "reliability" refers to the consistency of a test score: if a test is reliable it will yield consistent results over time and without regard to the irrelevant conditions such as the person administering the test. Internal consistency is considered an aspect of reliability: it means that all the items in a test measure the same thing. Temporal reliability is also called "test-retest reliability" because it is typically evaluated by having groups of individuals take the same test on several occasions and seeing how their scores compare Some differences are expected because of the random nature of the error component, but there should be a strong relationship between the observed scores of individuals on multiple occasions.

The term "inter-rater reliability" refers to the consistency of a test or scale regardless of who administers it. For instance, psychiatric conditions are often evaluated by having an observer rate an individual's behavior using a scale, and the results for different observers evaluating the same individual at the same time should be similar. For instance, three psychologists using a scale to evaluate the same child for hyperactivity should reach similar conclusions. Both types of reliability are typically evaluated by correlating test results on different occasions (temporal) or the scores returned by different raters (inter-rater).

Internal consistency can be measured in several ways. The split-half method involves having a group of individuals take a test, then splitting the items into two groups (for instance, odd numbered items in one group and even in the other) and calculating the correlation between the total scores of the two groups. Cronbach's alpha (or coefficient alpha) is a refinement of the split-half method: it is the mean of all possible split-half coefficients. The measure was developed and named "alpha" by Lee Chronbach, an educational psychologist and measure theorist who began his career as a high school mathematics and chemistry teacher.

The term "validity" refers to whether a test measures what it claims to be measuring. Three types of validity are typically discussed: content, predictive, and construct. Content validity refers to whether the test includes a reasonable sample of the subject or quality it is intended to measure (for instance, mathematical aptitude or quality of life) and is usually established by

Most psychological tests try to translate unobservable qualities such as intelligence or anxiety in terms of numbers. (Photos.com)

having a panel of experts evaluate the test in relation to its purpose. Predictive validity means that test scores correlate highly with measures of similar outcomes in the future; for instance, a test of mechanical aptitude should correlate with a new hire's success working as an auto repairman. Construct validity refers to a pattern of correlations predicted by the theory behind the quantity being measured: the scores on a test should correlate highly with scores on other tests that measure similar qualities and less highly with those that measure different qualities.

Further Reading

Embretson, Susan E., and Steven P. Reise. *Item Response Theory for Psychologists*. Mahwah, NJ: Erlbaum, 2000.

Furr, R. Michael, and Verne R. Bacharach. *Psychometrics: An Introduction*. Thousand Oaks, CA: Sage, 2007.

Gopaul McNicol, Sharon-Ann, and Eleanor Armour-Thomas. *Assessment and Culture: Psychological Tests with Minority Populations*. Burlington, MA: Elsevier, 2001.

Kline, Paul. *The Handbook of Psychological Testing*. 2nd ed. New York: Routledge, 2000.

Wood, James M., Howard N. Garb, and M. Teresa Neszworski. "Psychometrics: Better Measurement Makes Better Clinicians." In *The Great Ideas of Clinical Science: 17 Principles That Every Mental Health Professional Should Understand*. Edited by Scott O. Lilienfeld and William T. O'Donohue. New York: Routledge, 2007.

Sarah Boslaugh

Radiation

Category: Weather, Nature, and Environment.
Fields of Study: Algebra; Data Analysis and Probability; Measurement; Number and Operations.
Summary: Radiation research has a heavy mathematical component, especially in modeling distribution of or shielding from radiation.

Radiation is the transmission of energy via waves or particles, such as energetic electrons, photons, or nuclear particles. These waves or particles, called "quanta," travel radially in all directions from the source, leading to the name "radiation." Radiation exists everywhere, from both natural sources, like the sun, and many man-made sources, like radio stations and particle accelerators. The various types of radiation that exist may be harmful or beneficial to people, depending on source and application. Ionizing radiation contains enough energy per quantum to detach electrons from atoms, like X-rays or the radiation emitted by particle accelerators. High energy particles are created constantly by all luminous objects in the universe. Most of these particles never reach the surface of Earth. They may be deflected by magnetic fields or interact with atmospheric particles. Common types of nonionizing radiation include visible light, radio waves, and microwaves.

Many mathematicians have contributed to radiation research, like Wilhelm Wien, who derived a distribution law of radiation and won a Nobel Prize for his work on heat radiation. Physicist Max Planck used some of Wein's mathematics as the basis for quantum theory. Paul Ehrenfest contributed to quantum statistics, in part by applying Plank's quantum theory to rotating bodies. Subrahmanyan Chandrasekhar won the Royal Society Copley Medal for his work in mathematical astronomy, including the theory of radiation. Victor Twersky was widely regarded as an expert on radiation scattering. His work has been used in diverse applications, such as studying the effect of atmospheric dust on light propagation. Mathematicians continue to work on radiation problems, including applications such as detecting radiation or shielding satellites from the harmful effects of cosmic radiation, as well as creating mathematical methods for formulating and investigating radiation problems, such as Monte Carlo simulations.

Properties

Properties of radiation waves can be used to determine their potential effects on people and objects or their

Electromagnetic Radiation

Electromagnetic radiation (EMR) includes both ionizing and nonionizing forms of radiation. EMR waves result from the coupling of an electric field and a magnetic field. The fields are perpendicular to one other and to the direction of energy propagation. Electromagnetic radiation behaves like both a wave—with properties including reflection, refraction, diffraction, and interference—and a particle, because its energy occurs in discrete packets or quanta. Maxwell's equations, named for physicist and mathematician James Maxwell, are cited as the most elegant way to express the fundamentals of electromagnetism. The set of four equations, which have integral and differential forms include: Gauss's laws for electricity and magnetism, named for mathematician Carl Freidrich Gauss; Faraday's law of induction, named for physicist and chemist Michael Faraday; and Ampere's law with Maxwell correction, named for physicist and mathematician Andre-Marie Ampere. Many have derived theories and applications from these building blocks, such as mathematician Josef Stefan, who showed that total radiation from a blackbody is proportional to the fourth power of its absolute temperature.

usefulness for applications. Wavelength is the length of one cycle of the wave, or the distance from one peak to the next. Frequency is the number of cycles of the wave that travel past a fixed point along its path per unit time. All electromagnetic waves travel in a vacuum at a speed of about 3×10^8 meters per second. A fundamental relationship between wavelength and frequency is that wave speed is the product of wavelength and frequency, which means that greater wavelengths correspond to lower frequencies. The energy of electromagnetic photons is the product of wave frequency and Planck's constant, so higher frequencies produce greater photon energies. Among the common types of EMR radiation, radio waves have the longest wavelengths, resulting in low frequencies and low energies. Higher frequency ultraviolet radiation has the most energy and is the most harmful component of the cosmic radiation that penetrates Earth's atmosphere. X-rays, discovered by physicist Wilhelm Röntgen, occur naturally when solar wind is trapped by Earth's magnetic field in the Van Allen belts, named for physicist James Van Allen.

Black holes are also sources of X-rays in the universe. While photons have no mass, some forms of radiation are particles with positive mass produced in the atomic decay of radioactive materials. For example, beta radiation is composed of high-energy electrons, which are dangerous because they can penetrate skin to the layer where new cells are produced. Mathematician Jesse Wilkins's work on mathematical models to compute the penetration and absorption of electromagnetic gamma rays has been used in the design of nuclear radiation shields.

Further Reading

Dupree, Stephen, and Stanley Fraley. *A Monte Carlo Primer: A Practical Approach to Radiation Transport.* New York: Springer, 2001.

Knoll, Glenn. *Radiation Detection and Measurement.* Hoboken, NJ: Wiley, 2010.

U.S. Environmental Protection Agency. "Radiation Protection." http://www.epa.gov/radiation/programs.html.

Sarah J. Greenwald
Jill E. Thomley

Recycling

Category: Weather, Nature, and Environment.
Fields of Study: Algebra; Data Analysis and Probability; Number and Operations.
Summary: Efficient recycling requires the use of sophisticated mathematical models to maximize product use and reuse and minimize energy consumption.

Recycling is the extraction of usable materials out of used objects. Materials that are often recycled at the start of the twenty-first century include metal, paper, glass, and plastic. One important mathematical problem of recycling is the comparison of environmental and monetary costs of recycling and virgin production. Mathematicians are also involved in developing new methods for recycling and modeling both economic and environmental impacts. The notion of "algorithm recycling" applies to resources used in some mathematical investigations. For example, statistical bootstrap recycling reuses samples to minimize demands on computational resources. Some mathematicians, scientists, educators, and others use recycling for education, recreation, and art. Mario Marin has designed polyhedral outdoor play spaces and kinetic sculptures from recycled and remaindered materials and has published many creative ways to recycle household objects, like plastic bottles, into interesting polyhedral structures. With regard to learning, some have even suggested a concept called "neuronal recycling," which refers to adaption of neuronal circuits for new uses.

Proportion-Based Regulations and Labeling

To motivate recycling, companies and governments set rules that demand the recycling of a certain proportion of materials and the use of a certain proportion of recycled material in production. Because recycling is the third desirable option in the waste management hierarchy, after reduction of waste and reusing of objects and materials, setting high recycling quotas is never a goal in its own right. However, recycling is often preferable to disposal. Governments sometimes directly mandate minimum recycled content in certain classes of manufactured goods. Labeling laws, which require companies to display the percent of recycled content in goods and packages, may also promote recycling if consumers support it, or hinder recycling if consumers do not find recycled goods in this particular industry appealing. Companies advertise their recycling efforts—typically by disclosing the percent of recycled material—to present ecofriendly images to their customers.

A common scheme to promote the recycling of packaging is to include a refundable fee in the price of the product. Once the customer returns the packaging to the store, the fee is refunded.

Measuring Efficiency

Because recycling is a complex process, there are ecological and economical costs involved in it. For recycling to make sense, the benefits have to outweigh the costs. Computing costs and benefits is a complex problem. Costs are incurred at all stages of recycling: collecting materials, sorting them, and re-making them. Benefits include the reduction of landfill costs, reduction of pollution, and revenues from the use of recycled materials. In the cases of nonrenewable natural resources, recycling is the only option to keep using these resources in the future.

Metal Recycling

Because of the relative difficulty and high cost of mining and smelting of metals, and the ease of collecting and recycling, metals are the most recycled materials in the world. For example, recycling aluminum takes only 5% of the energy that it would take to make it from the raw materials. About three-quarters of steel and a third of aluminum is recycled in the United States as of 2010. Some applications of science and mathematics metal recycling involves the separation of impurities, such as paint.

Paper Recycling

One category of paper recycling, post-consumer paper, is familiar to most people because paper is ubiquitous in modern society. "Mill broke" is scraps that pulp mills accumulate from making paper, which they can also recycle. Preconsumer paper is scraps collected and recycled in paper mills. Unlike metal recycling, where the cost-benefit ratio is low, paper recycling is more complicated and controversial. For example, burning paper for energy may be more environmentally sound than recycling it and harvesting and replanting forests may be cheaper than recycling.

Estimates for energy saving are 40% to 65% for recycled paper, compared to creating new paper. However, pulp mills frequently produce energy by burning roots, bark, and other byproducts, whereas recycling plants have to be close enough to collection (usually urban) areas to minimize transport cost and frequently depend on fossil fuels for energy. Thus, the environmental costs of conserving the same amount of energy is different, as one process uses renewable resources and the other uses nonrenewable resources. Water and air pollution benefits of paper recycling are more pronounced than energy benefits because of highly toxic bleaching used in making new paper.

Plastic Recycling

Recycling of plastics involves a scientific challenge not found in recycling of other materials. Because of the ways polymer chains are formed in plastics, different plastics do not blend well. Removing dyes, glue, paper stickers, and other impurities is also difficult. Plastics are coded with the Resin Identification Codes, numbers 1–7, inside the triangular recycle symbol.

There are several processes for recycling plastic. The most straightforward is melting similar plastics together, with some steps to remove impurities. Heat compression mixes all types of plastics in high-heat, high-pressure drums. Thermal depolymerization is currently an experimental procedure that "reverses" the process of making plastic and turns it into a substance similar to crude oil. Another experimental procedure, called "monomer recycling," reverses plastic-making halfway, turning polymers into the mix of monomer chemicals that formed them.

The short-term cost-benefit analysis may not support plastic recycling because of the high energy and labor requirements of the known processes. However, crude oil (the raw material of plastic) is a nonrenewable resource, which makes plastic recycling attractive in the long term.

Glass Recycling

The main benefits of glass recycling are saving landfill space and saving energy on producing new glass. However, because glass is sturdy and easy to clean, glass container reuse is vastly preferable to recycling. Through changing their infrastructures, along with using clear bottle standards and monetary incentives, some countries can reuse more than 95% of their glass bottles.

Crushed glass can be added to concrete. This process can be considered reuse rather than recycling because the glass is serving a different purpose. Measurements of glass-infused concrete include its insulation properties and strength properties, both of which are improved by the addition of glass. Also, concrete with glass is more aesthetically pleasing and can be used for countertops and other highly visible places.

Mathematical Modeling

Mathematical models are widely used in logistics—controlling the efficient flow and storage of goods, services, and information from the point of origin to the point of consumption. Reverse logistics is the extension of this principle that addresses concepts such as returns, source reduction, recycling, and reuse. Mathematicians have researched models for logistics that address these reversals of flows. For example, Italian researchers created a staged mathematical model of the options for recycling a broad range of appliances, electronic equipment, and other household items commonly thrown away. The model suggested that recycling can offer what is known as *economies of scale* to businesses, which are increasingly being held liable for end-of-life product disposal.

Others have used techniques such as dynamic quantitative models to simulate recycling systems and flows to better understand the driving variables and relationships among the activities and participants. These models can aid planners in making decisions about recycling policies and procedures. Nutrient recycling for trees, which has implications for issues such as global warming, has been modeled using linear and quadratic functions, along with data-based numerical simulations. However, some scientists argue that mathematical models must be contextually evaluated and used with caution for decision making and legislation. Models based on limited data may generate what appear to be useful results, but extrapolation or subsequent modeling can create bias and propagation of errors.

Further Reading

Environmental Protection Agency. "Wastes—Resource Conservation—Reduce, Reuse, Recycle." http://www.epa.gov/osw/conserve/rrr/recycle.htm

La Mantia, Francesco. *Handbook of Plastics Recycling.* Shropshire, England: Smithers Rapra Technology, 2002.

Mancini, Candice. *Garbage and Recycling (Global Viewpoints).* Farmington Hills, MI: Greenhaven Press, 2010.

Newton, Michael, and Charles Geyer. "Bootstrap Recycling: A Monte Carlo Alternative to the Nested Bootstrap." *Journal of the American Statistical Association* 89, no. 427 (1994).

Schlesinger, Mark. *Aluminum Recycling.* Oxfordshire, England: Taylor & Francis Group, 2007.

Maria Droujkova

Relativity

Category: Space, Time, and Distance.
Fields of Study: Algebra; Geometry; Measurement; Representations.
Summary: Albert Einstein's theory of relativity is one of the most well-known theories in physics and helps describe the nature of the universe.

Albert Einstein's theory of relativity forms one of the two pillars of modern physics, the other being quantum mechanics. It consists of two parts: the special theory of relativity from 1905, and the general theory of relativity from 1915, which both rely on significant mathematics.

The special theory of relativity describes how space and time are perceived by observers in different inertial systems. Einstein derived this theory from a single physical principle of relativity. It was discovered in 1632 by Galileo Galilei that the laws of mechanics are the same in all inertial systems—a discovery, known as "Galileo's principle of relativity," that constituted a radical break with the prevailing Aristotelian physics. Einstein's principle of relativity generalized this concept to all laws of nature, including Maxwell's laws of electromagnetism, which govern the propagation of light. It thus follows from Einstein's principle of relativity that the speed of light is the same in all inertial systems, a central result in the theory of relativity. Prior to Einstein, it was believed that light propagates through a luminiferous aether in the same way as sound propagated through air, but all attempts to measure the speed of the Earth relative to this aether, such as the Michelson–Morley experiment in 1887, failed. Special relativity explained the negative results of these experiments and made the aether hypothesis superfluous.

The general theory of relativity unifies special relativity with Isaac Newton's law of universal gravity. Its basis is Einstein's equivalence principle, according to which an accelerated system of reference (such as a so-called Einstein elevator) is indistinguishable from a system at rest in a gravitational field. Mathematically, Einstein's field equations describe how the presence of mass, energy, and momentum gives rise to a curvature of space and time. Although this idea has little significance in weak gravitational fields, such as that of the Earth, general relativity is essential in the study of the universe as a whole. For example, Karl Schwarzschild in 1915 found an exact solution to Einstein's equations that explains the existence of black holes.

The many surprising consequences of the theory of relativity have been described in numerous popularizations, most notably by George Gamow. Einstein's theory must not be confused with the various relativist positions in philosophy, such as aesthetic, moral, cultural, or cognitive relativism.

Special Relativity

The Lorentz transformation forms the basis of the special theory of relativity. It is a set of equations describing how to translate suitably chosen coordinates of space and time between two inertial systems (S) and (S') moving with the speed (v) relative to one another:

$$x' = \gamma(x - vt) \text{ and } t' = \gamma\left(t - \frac{vx}{c^2}\right)$$

where c denotes the speed of light of 299,792,458 meters per second, and the dimensionless number

$$\gamma = \frac{1}{\sqrt{1 - \frac{v^2}{c^2}}}$$

is the so-called Lorentz factor. In 1908, Hermann Minkowski gave a mathematical description of the Lorentz transformation as a rotation of the coordinate axes in four-dimensional space-time.

When v is much smaller than c, the Lorentz factor is close to 1, and the Lorentz transformation reduces to the classical Galilean transformation. When v approaches c, however, the Lorentz transformation has a number of consequences that radically contradict classical physics as well as common sense. For example, clocks in motion are slowed down (called "relativistic time dilation"), objects in motion are contracted in the direction of movement (called "relativistic length contraction"), and clocks in motion that are seen as synchronized by an observer moving with the clocks are seen as nonsynchronized by an observer at rest (called "relativity of simultaneity").

It is another consequence of special relativity that no material objects—or signals of any kind—can travel faster than light. This "speed limit" exists because anything traveling faster than light relative to one observer would appear to be traveling backwards in time relative to another observer, thus leading to paradoxes regarding cause and effect. There is a quantum-mechanical phenomenon, the so-called Einstein–Podolsky–Rosen paradox, that seems to contradict this principle. According to quantum mechanics, the wave function of two entangled particles is affected by a measurement of the state of one of the particles, causing an instantaneous change to the state of the other, even if the two particles are located in different galaxies. But this phenomenon, which has since been verified experimentally, does not really contradict relativity since it cannot be used to transmit information from one galaxy to the other.

Special relativity dictates that mass and energy are connected by the equation $E = mc^2$, undoubtedly the most famous formula in all of physics. Any particle with mass m has a rest energy given by this equation. If the same particle is accelerated to the speed v, its energy is multiplied by the Lorentz factor γ, and its kinetic energy is found as the difference between total energy and rest energy, expressed algebraically as

$$E_{kin} = \gamma mc^2 - mc^2 \approx \frac{1}{2}mv^2.$$

The approximation, valid for v much smaller than c, equals the expression for kinetic energy in classical mechanics. This formula shows that it would require an infinite amount of energy to accelerate a particle with positive mass to the speed of light.

General Relativity

Einstein noted that special relativity implies that space appears to be curved, or "non-Euclidean," to observers in accelerated systems (for example, on a rotating disc) and inferred from the equivalence principle that the same must be true in gravitational fields. However, after realizing this fundamental principle in 1907, it took him eight years to find the field equations that describe the exact curvature of space-time. The idea that physical space might be curved was not new. Already in 1823, Carl Friedrich Gauss investigated this question empirically by measuring the sum of angles of a triangle formed by three mountaintops but found no curvature. Bernhard Riemann further developed the mathematics of curved space in 1854 and this work would become an essential part of Einstein's theory.

General relativity predicts that a body falling freely in a gravitational field, such as the Earth in its orbit around the sun, follows a "geodesic" in curved space-time. This geodesic is called the body's "world-line." In a curved space, geodesics are the least curved lines, in the same way as the equator is a least curved line on the surface of Earth. Although the predictions of general relativity are nearly the same as those of classical mechanics for bodies in weak gravitational fields, the interpretation of gravity is radically different: whereas classical mechanics explains the elliptical orbit of the Earth as a consequence of a gravitational force emanating from the sun, general relativity postulates that the mass of the sun gives rise to a curvature of space-time, and that the world-line of Earth is in fact a geodesic.

It is a consequence of general relativity that clocks in gravitational fields are slowed down. This effect is called "gravitational time dilation." For a clock at rest in the gravitational field of Earth, the dilation factor is

$$\sqrt{1 - \frac{2GM}{rc^2}} \approx 1 - \frac{GM}{rc^2}$$

where G is Newton's gravitational constant, M is the mass of Earth, and r is the distance between the clock and the center of Earth.

Proofs and Applications of Relativity

Einstein showed in 1915 that general relativity explains the perihelion precession of the planet Mercury. This phenomenon, which had mystified astronomers since its discovery in 1859, is that the elliptical orbit of Mercury rotates around the sun with 43 arc seconds per century.

Also in 1915, Einstein predicted that light emitted from distant stars is deflected when passing through the gravitational field of the sun. Although this effect had previously been derived from Newtonian gravity alone, Einstein showed that the angle of deflection following from general relativity is twice the angle following from classical physics. Einstein's prediction was confirmed dramatically by Arthur Eddington during the total solar eclipse of May 29, 1919.

Contrary to quantum mechanics, the technological implementations of which are ubiquitous, relativity has few practical applications. One notable exception is the global positioning system (GPS). GPS satellites revolve around the Earth twice per sidereal day at a height of about 20,000 kilometers (12,400 miles) and with a speed of about 4 kilometers (2.5 miles) per second. Because of the speed and altitude, the atomic clocks aboard the satellites are subject both to relativistic time dilation and to a reduced gravitational time dilation.

The first effect amounts to a loss of 7 microseconds per day, the second to a gain of 45 microseconds per day. In total, therefore, the atomic satellite clocks gain 38 microseconds per day relative to clocks on the ground. Failure to take these relativistic effects into account would render GPS useless since the resulting positional error would accumulate to 11 kilometers (6.8 miles) per day.

Further Reading

Einstein, Albert. *Relativity: The Special and General Theory.* New York: Henry Holt, 1920.

Feynman, Richard, Robert Leighton, and Matthew Sands. *The Feynman Lectures on Physics.* Reading, PA: Addison-Wesley, 1964.

Gamow, George. *Mr. Tompkins in Wonderland.* New York: Macmillan, 1946.

Grøn, Oyvind, and Sigbjorn Herv. *Einstein's General Theory of Relativity: With Modern Applications in Cosmology.* New York: Springer Science+Business Media, 2007.

Møller, Christian. *The Theory of Relativity.* Oxford, Egland: Oxford University Press, 1952.

Russell, Bertrand. *The ABC of Relativity.* London: Kegan Paul, Trench, Trubner, 1925.

Ungar, Abraham. *Analytic Hyperbolic Geometry and Albert Einstein's Special Theory of Relativity.* Singapore: World Scientific Publishing, 2008.

David Brink

Satellites

Category: Communication and Computers.
Fields of Study: Algebra; Geometry; Measurement.
Summary: Mathematics is fundamental to the design, function, and launch of satellites.

Astronomy and mathematics have long developed together. Many early mathematicians studied the motion of celestial objects. The term "satellite" comes from the Latin *satelles* (meaning "companion"), which was used by mathematician and astronomer Johannes Kepler to describe the moons of Jupiter in the seventeenth century. Mathematician Giovanni Cassini correctly inferred that Saturn's rings were composed of many small satellites in the seventeenth century. Mathematicians Jean Delambre and Cassini Jacques both published books of astronomical tables, including planetary satellites, in the eighteenth century. When artificial satellites were developed, the term "satellite" largely came to refer to those in common speech, while "moon" was applied to natural bodies orbiting planets. Mathematicians like Michael Lighthill and engineers like John Pierce helped develop satellites in the 1960s.

By the first decade of the twenty-first century, there were several hundred operational satellites orbiting the Earth to facilitate communication, weather observation, research, and observation. The advantage of satellites for communication are that signals are not blocked by land features in the same manner as a lower-altitude signal would be, making long-distance communication possible without multiple ground-based relays. Early communication satellites simply reflected signals back to Earth to broaden reception. Modern satellites use many different kinds of orbits to facilitate complex functioning, including low Earth orbit; medium Earth orbit; geosynchronous orbit; highly elliptical orbit; and Lagrangian point orbit, named for mathematician Joseph Lagrange. Mathematics is involved in the creation and function of such satellites, as well as for solving problems related to launching satellites, guiding movable satellites, powering satellite systems, and protecting satellites from radiation in the Van Allen belt, named for physicist James Van Allen. For example, graph theory is useful in comparing satellite communication networks. Techniques of origami map folding, researched by mathematicians like Koryo Miura, have been used in satellite design. Chaos theory has been used to design highly fuel-efficient orbits, derived in part from mathematician Henri Poincaré's work in stable and unstable manifolds. Government agencies like the U.S. National Aeronautics and Space Administration (NASA) and private companies like GeoEye employ mathematicians for research and applications. The Union of Concerned Scientists (UCS) maintains a database of operational satellites.

Antennas and satellite dishes generally have a parabolic shape and are used to receive satellite signals on Earth. (Photos.com)

Orbits

The orbit of a satellite about the Earth determines when it will pass over various points on the Earth's surface and how high it is above the Earth. In general, orbits are characterized by altitude, inclination, eccentricity, and synchronicity. As defined by NASA, low Earth orbits have altitudes of 80–2000 kilometers. This orbit includes the majority of satellites, the International Space Station, and the Hubble Space Telescope. Statistical estimates at the start of the twenty-first century suggest that the number of functional satellites and nonfunctional debris in low orbit ranges from a few thousand (tracked by the U.S. Joint Space Operations Center) to millions (including very small objects). Objects in low orbit must travel at speeds of several thousand kilometers per hour, so even a small object can cause damage in a collision. Medium Earth orbit extends to about 35,000 kilometers (21,000 miles), the altitude determined by Kepler's laws of planetary motion for geosynchronous orbits. Inclination is an angular measure with respect to the equator, while eccentricity refers to how elliptical an orbit is. Geosynchronous satellites rotate at the same rate as the Earth spins, so they appear stationary relative to Earth. They usually have inclination and eccentricity of zero; they circle the equator to balance gravitational forces. The Global Positioning System (GPS) is one example of satellites at this orbital level. Sun synchronous orbits are retrograde patterns that allow a satellite to pass over a section of the Earth at the same time every day. They have an inclination of 20–90 degrees and must shift by approximately one degree per day. These orbits are often used for satellites that require constant sunlight or darkness. The maximal inclination of 90 degrees denotes a polar orbit. A halo or Lagrangian orbit is a periodic, three-dimensional orbit near one of the Lagrange points in the three-body problem of orbital mechanics, which was used for the International Sun/Earth Explorer 3 (ISEE-3) satellite.

Signals

Antennas and satellite dishes are used to receive satellite signals on Earth. Most satellite dishes have a parabolic shape. A signal striking a planar surface reflects directly back to the source. If the surface is curved, the reflection is in the plane tangent to the surface. A parabola is the locus of points equidistant from a fixed point and a plane, so a parabolic dish focuses all incoming signals to the same point at the same time, increasing the quality of the signal. Mathematics is used to compress, filter, interpret, and model vast amounts of data produced by satellites. Reed–Solomon codes, derived by mathematicians Irving Reed and Gustave Solomon, are widely used in digital storage and communication for satellites. Much of the data from satellites is images, which utilize mathematical algorithms for rendering and restoration. One notable case that necessitated mathematical correction is the Hubble Space Telescope. An incorrectly ground mirror was found to have a spherical aberration, which resulted in improperly focused images. Mathematical image analysis allowed scientists to deduce the degree of correction needed. Some of the mathematical concepts involved in these corrections include the Nyquist frequency, which is a function of the sampling frequency of a discrete signal system named for physicist Harry Nyquist, and the Strehl ratio, named for mathematician Karl Strehl, which quantifies optical quality as a fraction of a system's theoretical peak intensity.

Further Reading

Montenbruck, Oliver, and Gill Eberhard. *Satellite Orbits: Models, Methods and Applications.* Berlin: Springer, 2000.

Whiting, Jim. *John R. Pierce: Pioneer in Satellite Communication.* Hockessin, DE: Mitchell Lane Publishers, 2003.

BILL KTE'PI

Solar Panels

Category: Architecture and Engineering.
Fields of Study: Algebra; Geometry; Measurement.
Summary: The angle of inclination of a solar panel array is key to its efficiency, among other factors.

Solar panels are interconnected assemblies of photovoltaic cells that collect solar energy as part of a solar power system, either on Earth or in space. Typically, several solar panels will be used together in a photovoltaic array along with an inverter and batteries to store collected energy. Photovoltaic cells convert the energy of sunlight into electricity via the photovoltaic effect (the creation of electric current in a material when it is exposed to electromagnetic radiation), which was observed by French physicist Alexandre-Edmond Becquerel in 1839. Prior to that time, many scientists and mathematicians built and researched parabolic burning mirrors, which are another way to focus solar energy. Diocles of Carystus showed that a parabola will focus the rays of the sun most efficiently. Archimedes of Syracuse may have built burning mirrors that set ships on fire. George LeClerc, Comte de Buffon, apparently tested the feasibility of such a mirror by using 168 adjustable mirrors in order to vary the focal length to

The solar field at Nellis Air Base in Nevada has more than 72,000 panels and supplies the base with more than 30 million kilowatt-hours of power. (U.S. Air Force)

ignite objects that were 150 feet away. It was also investigated experimentally in the early twenty-first century on the television program *Mythbusters*. Mathematics teacher Augustin Mouchot investigated solar energy in the nineteenth century and designed a steam engine that ran on sun rays. Some consider this invention to be the start of solar energy history. The first working solar cells were built by the American inventor Charles Fritts, in 1883, using selenium with a very thin layer of gold. The energy loss of Fritts's cells was enormous—less than 1% of the energy was successfully converted to electricity—but they demonstrated the viability of light as an energy source. Engineer Russell Ohl's semiconductor research led to a patent for what are considered the first modern solar cells, and Daryl Chapin, Calvin Fuller, and Gerald Pearson, working at Bell Labs in the 1950s, developed the silicon-based Bell solar battery. There were fewer than a single watt of solar cells worldwide capable of running electrical equipment at that time. Roughly 50 years later, solar panels generated a billion watts of electricity to power technology on Earth, satellites, and space probes headed to the far reaches of the galaxy. Scientists and mathematicians continue to collaborate to improve solar panel technology. One such focus is creating scalable systems that are increasingly efficient and economically competitive with various other energy technologies.

Physics and Mathematics of Solar Panels

In 1905, Albert Einstein published both a paper on the photoelectric effect and a paper on his theory of relativity. His mathematical description of photons (or "light quanta") and the way in which they produce the photoelectric effect earned him the Nobel Prize in Physics in 1921. In general, the photons or light particles in sunlight that are absorbed by semiconducting materials in the solar panel transfer energy to electrons—though some is lost in other forms, such as heat. Added energy causes the electrons to break free of atoms and move through the semiconductor. Solar cells are constructed so that the electrons can move in only one direction, producing electrical flow. A solar panel or array of connected solar panels produces direct current, like chemical batteries, which can be stored. An inverter can convert the direct current to alternating current for household use.

Mathematics is involved in many aspects of solar panel design, operation, and installation. For example, the perimeter of an array of multiple solar panels may change with rearrangement of the panels, but the area stays the same. Since area is one critical variable for power collection, this suggests different optimal arrangements for surfaces where solar panels might be arranged, like walls and roofs. Satellites often use folding arrays of solar panels that deploy after launch, and folding portable solar panel arrays have been designed for applications like camping and remote or automated research and monitoring stations. Space scientist Koryo Miura developed the Muria–Ori map folding technique, which involves mathematical ideas of flexible polygonal structures and tessellations. It has been incorporated into satellite solar panels that can be unfolded into a rectangular shape by pulling on only one corner.

Arrays

A solar panel array may be fixed, adjustable, or tracking. Each method has trade-offs in installation cost versus efficiency and energy over the lifetime of the installation, which can be analyzed mathematically in order to optimize an individual setup. Fixed arrays are solar panels that stay in one position. Optimal positioning of such arrays usually involves facing the equator (true south, not magnetic south, when in the northern hemisphere), with an angle of inclination roughly equal to their latitude. Using an angle of inclination slightly higher than the latitude has been shown in some studies to improve energy collection in the winter, which can help balance shorter days or increased heating energy needs. Setting the inclination slightly less than the latitude optimizes collection for the summer. Adjustable panels can have their tilt manually adjusted throughout the year. Tracking panels follow the path of the sun during the day, on either one or two axes: a single-axis tracker tracks the sun east to west only, while a double-axis tracker also adjusts for the seasonal declination movement of the sun. Tracking panels may lead to a gain in power, but for some users, the cost trade-off might suggest adding additional fixed panels for some applications instead. Solar power companies and other entities provide maps showing the yearly average daily sunshine in kilowatt hours per square meter of solar panel. Combined with the expected energy consumption of a building, this data helps determine how many solar panels and batteries will be needed for an installation. Science and mathematics teachers often have students build solar panels and collect data to facilitate mathematical understanding and critical thinking, as well as make mathematics, science, and technology connections.

Further Reading

Anderson, E. E. *Fundamentals of Solar Energy Conversion.* Reading, MA: Addison Wesley Longman, 1982.

Hull, Thomas. "In Search of a Practical Map Fold." *Math Horizons* 9 (February 2002).

Kryza, F. *The Power of Light: The Epic Story of Man's Quest to Harness the Sun.* New York: McGraw-Hill, 2003.

Bill Kte'pi

Stethoscopes

Category: Medicine and Health.
Fields of Study: Algebra; Geometry.
Summary: Some modern stethoscope designs digitize sound waves, which can be modeled and analyzed.

The stethoscope is perhaps one of the most iconic pieces of medical equipment and is used by doctors in nearly every area of clinical practice around the world. From its beginnings as a simple tube to amplify sound, in the twenty-first century the stethoscope is evolving into a highly mathematical and computerized tool. It can record, analyze, and display diagnostic information using software and algorithms developed from clinical data using a variety of concepts and techniques from statistics, signal processing, spectral analysis, and related sciences. Further, mathematical models and simulations are increasingly used to support and validate clinical results.

History and Development

French physician René Laennec is credited with the invention of the "stethoscope" in 1816. The name comes from the Greek words meaning "chest" and "to examine." Knowing that solid bodies conduct and amplify sound, Laennec used tightly rolled and glued sheets of paper to hear patients' heartbeats. Experimenting with cylinders of various materials, he observed that an aperture maximized magnification of internal body sounds. His ultimate design was a straight, eight-inch wooden tube with a conical chest piece and a funnel-shaped stopper. Later physicians developed stethoscopes from materials like rosewood, papier-mâché, and even glass. The binaural form was popularized in the United States in the early 1900s by William Osler.

In the twenty-first century, the binaural acoustic stethoscope consists of a chest piece with a plastic disc (called a "diaphragm") on one side and a hollow cup (called a "bell") on the other. The bell transmits low frequency sounds and the diaphragm transmits high frequency sounds. A majority of clinicopathological correlations and diagnostic techniques used today result from patient data acquired by physicians listening with stethoscopes or a bare ear. Refinements in design and the increasingly widespread use of stethoscopes—coupled with training—improved observations. With respect to the heart, these included better precision in timing cardiovascular sounds, focusing on segments of the cardiac cycle in turn, and devising quantitative symbols to describe sounds. On the other hand, stethoscopes have also been investigated as a vector of disease transmission in busy clinical settings like emergency rooms.

Mathematical Modeling

Electronic systems of collecting and analyzing data have begun to supplement or even supplant the use of the stethoscope. Some predict that before 2020, manual stethoscopes will become obsolete. Electronic stethoscopes convert acoustic sound waves into electrical signals, which can be amplified and enhanced, producing both visual and audio output. Software can then represent cardiopulmonary sounds graphically and interpret them using mathematical algorithms. Signals may also be recorded or transmitted, facilitating remote diagnosis and teaching. Some research suggests that mathematical methods improve accuracy in diagnosing conditions, such as heart murmurs, but some methods have not yet shown clinical usefulness. Mathematicians and physicians continue to investigate and model cardiac sounds from murmurs and prosthetic valves, as well as other types of hemodynamic data, using techniques from spectral waveform analysis and physics concepts like damped oscillations of viscoelastic systems. They have also sought to quantify pulmonary sounds, like wheezing and crackles, and address signal processing issues, such as noise reduction, amplification, and filtration.

Measuring Blood Pressure

Blood pressure is the amount of pressure exerted by the blood upon the arterial walls. A clinician uses a device known as a "sphygmometer"—a device that pumps air into a cuff wrapped around a patient's arm—and listens for pulse sounds with a stethoscope, observing the height in millimeters of a column of mercury supported by the blood pressure. The sounds are known as "Korotkoff sounds," named for Russian physician Nikolai Korotkoff. A contraction of the heart that causes a pulse beat that supports a column of mercury 120 millimeters high is called a "systolic reading of 120." The reading in the period between contractions of the heart or pulses is called the "diastolic blood pressure." If the diastolic reading is 80 millimeters, the blood pressure is recorded as 120/80 and is read as "120 over 80." These numbers represent a ratio rather than a true fraction. The U.S. National Heart, Lung and Blood Institute defines normal blood pressure to be <120 for systolic *and* <80 for diastolic pressure and defines hypertension to be >140 *or* >90 for systolic and diastolic, respectively. These values are derived in part from statistical studies of typical human variation in blood pressure and associations with medical conditions like stroke and heart disease. Early diagnosis and appropriate treatment of hypertension is recognized as one of the most significant advances of modern medicine in reducing morbidity and mortality.

Further Reading

Bishop, P. J. "Evolution of the Stethoscope." *Journal of the Royal Society of Medicine* 73 (1980). http://www.ncbi.nlm.nih.gov/pmc/articles/PMC1437614.

Pullan, Andrew, Leo Cheng, and Martin Buist. *Mathematically Modeling the Electrical Activity of the Heart: From Cell to Body Surface and Back Again.* Singapore: World Scientific Publishing, 2005.

Karen Doyle Walton

Sunspots

Category: Weather, Nature, and Environment.
Fields of Study: Algebra; Data Analysis and Probability.
Summary: Sunspots have long been observed and mathematicians and scientists continue to try to understand them and their effects.

Sunspots are a not yet fully explained phenomenon tied to solar activity. The sun is Earth's richest source of heat and light. Furious eruptions of energy take place on the surface of the sun. In the core, nuclear reactions occur because of the immense temperature and pressure. Through a process known as "convection," millions of tons of hydrogen are converted into helium every second and are then expelled at the surface of the sun as light and heat. Sunspots have a magnetic field strength that is thousands of times stronger than Earth's magnetic field. These magnetic fields inhibit convection to create relatively cooler areas, which appear as dark spots on the surface of the sun. Scientists and mathematicians have long attempted to understand their behavior and oscillations and have used mathematical tools like differential equations, hexagonal planforms, and time series analyses. They also count the number of sunspots and examine possible relationships between this number and factors on Earth, like radio disruptions, land temperature, and weather phenomena.

History

Direct observation of the sun is very dangerous, which historically made sunspots hard to study and quantify. In ancient times, Chinese astronomers recorded solar activity. Mathematician and astronomer Thomas Harriot, noted for his work on algebra, is also credited as the discoverer of sunspots. Increased understanding of the nature of sunspots, including the observation that they often occurred in groups and that they moved relative to one another as the sun rotated, is tied to the development of the telescope in the seventeenth century. One of Galileo Galilei's works on sunspots offered evidence for the heliocentric system of Nicolaus Copernicus, and this led to debate about sunspots, as evidenced in astronomer, mathematician, and Jesuit Christoph Scheiner's views and works.

In the eighteenth century, Alexander Wilson used a geometric argument to show that sunspots were depressions. In the nineteenth century, pharmacist and amateur astronomer Heinrich Schwabe collected data on the periodicity of sunspots. Systematic observations, such as the approximately 11-year cycle, were made by Rudolph Wolf starting in 1848, who also measured the number of sunspots present on the surface of the sun. Wolf was primarily an astronomer but he also taught mathematics and physics. His observations were disputed by other astronomers, but his methods, which were based on statistical analyses, were eventually accepted as correct. Wolf's formula continues to be used in the twenty-first century as one of the sunspot indices. The International Sunspot Number is compiled worldwide by the Solar Influences Data Analysis Center in Belgium and by the U.S. National Oceanic and Atmospheric Administration. In the twenty-first century, sunspots are observed with solar telescopes, which use various filters, and specialized tools such as spectroscopes and spectrohelioscopes. Amateurs generally observe sunspots using projected images.

Waxing and Waning

Scientists know that the sun had a period of relative inactivity in the seventeenth century, which corresponds to a climatic period called the "Little Ice Age." Evidence suggests that similar periods existed in the distant past, which means there might be a connection between solar activity and terrestrial climate. The magnetic activity that accompanies the sunspots can change the ultraviolet and soft X-ray emission levels, affecting Earth's upper atmosphere. Some researchers have proposed that sunspots and solar activity are the main cause of global warming rather than carbon dioxide greenhouse gas emissions.

Further Reading

Izenman, Alan. "J R Wolf and the Zurich Sunspot Relative Numbers." *The Mathematical Intelligencer* 7 (1985). http://astro.ocis.temple.edu/~alan/WolfMathIntel.pdf.

National Aeronautics and Space Administration. "The Sunspot Cycle." http://solarscience.msfc.nasa.gov/SunspotCycle.shtml.

Spaceweather.com. "The Sunspot Number." http://spaceweather.com/glossary/sunspotnumber.html.

SIMONE GYORFI

Synchrony and Spontaneous Order

Category: Weather, Nature, and Environment.
Fields of Study: Algebra; Connections; Data Analysis and Probability.
Summary: The world is filled with examples of spontaneously emerging order.

Humans are familiar with order: people order homes by placing belongings in one place; people also watch football games with players who follow orders given by a quarterback who directs the play. There are many examples of order in nature. Birds and fish order themselves by flying in flocks and swimming in schools. How is order created in a complex system with many parts? Experience indicates that order emerges from the actions or directions of a leader, just as the quarterback is the leader of a football team. It is possible, however, for a system to be ordered without the help of a single leader—an attribute that occurs in a spontaneously ordered system. Systems have a global (group) level and a local (individual) level. A school of fish is made up of thousands of individual fish, and a laser is a collection of particles of light (photons) that are emitted from trillions of atoms. When a system is spontaneously ordered, the order occurs because of local level interactions without global level direction. Imagine a spontaneously ordered football team. The quarterback on this team does not need to direct or call a play. This team is able to organize and execute plays simply by communicating with each other (individually) as each play unfolds. There will likely never be a team like this, but spontaneously ordered phenomena are all around if one knows where to look.

When multiple events are ordered in time, the result is "synchrony." Without synchrony, life would be very different. People would not enjoy watching a football game with unsynchronized players who run in different directions after—or before—the ball is snapped. Many of the technological devices that people use, including GPS, cell phones, and lasers, rely on synchrony to work properly. Scientists have even published evidence of synchrony in cloud patterns. When spontaneous order occurs, the result is often synchrony. Mathematicians and statisticians are involved in the collection of data that help define important variables related to synchrony and fuel the development of theories and models, as well as the formulation of mathematical models to describe and explain synchrony. This work draws from many areas of mathematics, including logic, probability, decision theory, geometry, and statistics, as well as related scientific fields.

Examples of Synchrony and Spontaneous Order

In some regions of Southeast Asia, large numbers of male fireflies flash on and off at the same time, creating a spectacular array of synchronized lights. It is believed that the males are flashing in unison to attract females. Physiologically, these fireflies have an internal firing mechanism that can generate a rhythmic flashing sequence. Experiments with individual fireflies demonstrate that the timing of their flashes can be altered to mimic that of an external stimulus, which is flashing rhythmically. This suggests that synchronized firefly flashing is the result of a spontaneously ordered process. To test this hypothesis, mathematicians Renato Mirollo and Steven Strogatz created a simple mathematical model by using an equation to describe an individual firefly as a biological oscillator (just as a plucked guitar string is a mechanical oscillator). They coupled multiple, identical oscillators together to form a system. Their mathematical model is a system of coupled differential equations. Mirollo and Strogatz analyzed the system and proved that in almost all cases, no matter how many oscillators there are or how the oscillations are started, synchrony is the result.

Fish often travel in schools. One advantage of this behavior is to allow fish to better avoid predators by performing highly synchronized, evasive maneuvers. Experimental data suggests that schooling fish have a preferred distance, elevation, and orientation relative to their nearest neighbor. Scientists Andreas Huth and Christian Wissel have modeled fish schooling as a spontaneously ordered system. They assume that schooling originates not because of a particular fish directing the group's movements but because of simple behavioral rules for individual fish. Their assumptions include that each fish desires to be close (but not too close) to another fish, each fish moves according to its perception of the position and orientation of neighboring fish, and individual fish movement is random. Huth and Wissel tested different movement rules for their model since there are no data that supports specific movement rules for schooling fish. They used the data generated from computer simulations of their model to determine the average direction of movement as a group and the average angular deviation by individual fish from the group's direction, which is defined as the "polarization" of the school. The polarization is a way to quantify the synchrony of the school because the larger the polarization, the more disoriented the school is. Since polarization depends on the movement rules, they used polarization to find movement rules for which their model best simulated synchronized schooling.

A fluorescent light bulb consists of a long tube filled with an inert gas. The light that we observe originates from the atoms in the gas. Each atom has multiple electrons that exist at specific energy levels. Electricity forces electrons through the tube and these electrons collide with the atoms in the gas. The collision raises the energy level of the atom's electrons, which then spontaneously revert back to a state of lower energy. This loss of energy causes a light particle (photon) to be emitted and the light that we see is from the emission of millions upon millions of photons. The light from a fluorescent light bulb consists of many different wavelengths and is scattered in many directions. Alternatively, the light from a laser, which stands for "light amplification by stimulated emission of radiation," is highly synchronized with a single frequency, direction, and phase. The first laser was constructed in 1960, but in 1917, Albert Einstein developed the quantum physics that predicted how a laser is able to synchronize the photons. When lasers were invented no one knew what to use them for.

Today, laser light is used for everything from grocery store checkout scanners to eye surgery. Just as with fluorescent light, raising and lowering the energy levels of individual electrons generates the light from a laser. An external energy source (such as electricity) continually stimulates electrons and raises them from lower energy states to higher energy states. Initially, when the laser is turned on and some electrons spontaneously fall back to their lower energy states, the emitted photons move in random directions. But a laser has mirrors at both ends and the photons are trapped between the mirrors for a long period of time before they can escape. Furthermore, a laser is constructed so that the photons will perfectly synchronize and amplify a light wave with a specific frequency and direction while filtering out the other light waves. One of the mirrors allows some of the light to escape in the form of a laser beam, an example of synchrony that we encounter each day.

Further Reading

Camazine, Scott, Jean-Louis Deneubourg, Nigel R. Franks, James Sneyd, Guy Theraulaz, and Eric Bonabeau. *Self-Organization in Biological Systems*. Princeton, NJ: Princeton University Press, 2001.

Haken, Hermann. *The Science of Structure: Synergetics*. New York: Van Nostrand Reinhold, 1981.

Strogatz, Steven. *SYNC: The Emerging Science of Spontaneous Order*. New York: Hyperion, 2003.

John G. Alford

Telescopes

Category: Space, Time, and Distance.
Fields of Study: Algebra; Geometry; Number and Operations.
Summary: Image clarity in telescopes is achieved through extremely precise measurements and mathematics.

In 1608, the Dutch lensmaker Hans Lippershey applied for a patent on what was soon named a "telescope." It is not clear if Lippershey was the true inventor; at least two other Dutch lensmakers also claimed credit. The news of this new invention quickly spread. In 1609, Galileo Galilei in Italy started using telescopes to observe heavenly objects. Among other findings, he discovered the rotation of the sun, the phases of Venus, and the first four satellites of Jupiter. A mathematician as well as a physicist and astronomer, Galileo also used geometry to measure the heights of lunar mountains by determining how long they remained illuminated after the lunar sunset.

Other mathematicians and physicists helped develop the modern telescope. Isaac Newton determined that lenses acted like prisms in spreading out the spectrum of visible light (a phenomenon known as *chromatic aberration*). Newton and the mathematician James Gregory independently invented the reflecting telescope, which does not have this problem. Leonhard Euler made a mathematical analysis of chromatic aberration, and in England so-called achromatic lenses (a combination of two lenses that together bring light of different colors to a focus) were invented in the early eighteenth century.

Optics

A telescope is an optical device for seeing objects that are either far away, or very dim, or both. Consider a typical magnifying glass, as shown in Figure 1, which is a piece of glass or other transparent substance shaped so that both sides are sections of spheres. Light rays from an object (such as a candle) come to a focus on Screen 1. In other words, light rays from any given point on the candle converge onto a single point of Screen 1, forming an image. Screen 2 is at the wrong distance, meaning the light rays do not converge properly on Screen 2. Screen 1 is said to be "in focus," and Screen 2 is "out of focus."

Figure 1.

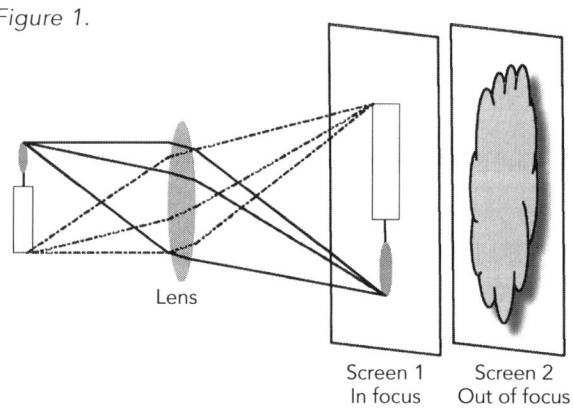

Screen 1 In focus   Screen 2 Out of focus

Figure 1a.

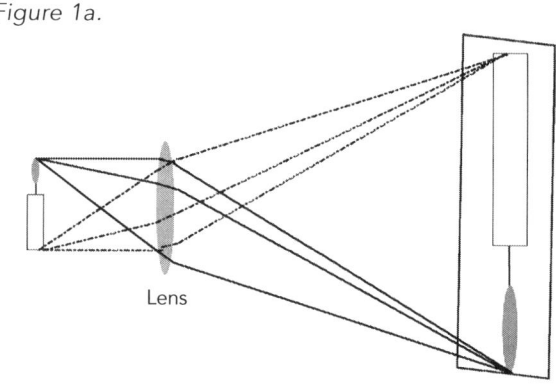

Figure 1 repeated, but with a lens with twice the focal length.

The first lens the light goes through is the called the "objective lens." The plane (Screen 1) where the image is in focus is called the "focal plane." The "focal length" is the distance to the focal plane for a source at infinity (incoming parallel rays).

Magnification is measured in diameters. If the image is twice as tall and also twice as wide as the original, then there is a magnification of two diameters. Some optical devices, however, are measured in power, which is the square of the magnification in diameters; for example, a microscope advertised as "100 power" actually magnifies 10 diameters.

Figure 1a shows the same configuration as Figure 1, but the focal length is twice as long. The image on the focal plane is thus twice as high and twice as wide—it is magnified twice as many diameters. Since the image is spread out over four times the area, it is only one-fourth as bright. Conversely, for a given focal length, doubling the size of the objective lens lets in four times as much light, hence the image is four times as bright.

In Figure 1, if one were to put a light-tight box around screen 1, set up a shutter to control when light enters the box, and replace screen 1 with photographic film, the result is a camera. Replace the photographic film with an electronic light-sensitive screen, and the result is a digital camera. If the camera is used to take pictures of far-away or dim objects, then it qualifies as a telescope.

Astronomical Telescopes

Since astronomers are interested in dim celestial objects, a big objective is necessary for astronomical telescopes. Amateur astronomers frequently use a 6-inch (15-cm) objective as a good compromise between light-gathering power and cost. Professional astronomers rarely use objectives less than about half a meter (1.5 feet) in diameter. The largest objective lens

Figure 2. Eye plus magnifying glass.

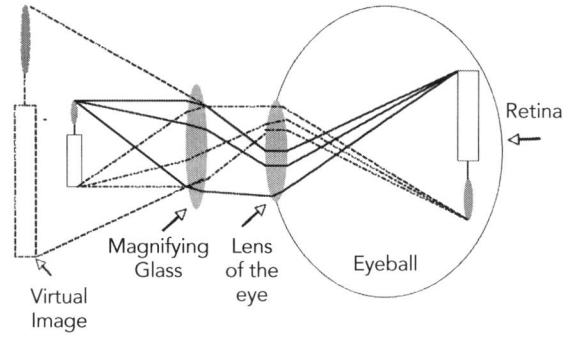

Figure 3. How a telescope delivers an enlarged image to the eye.

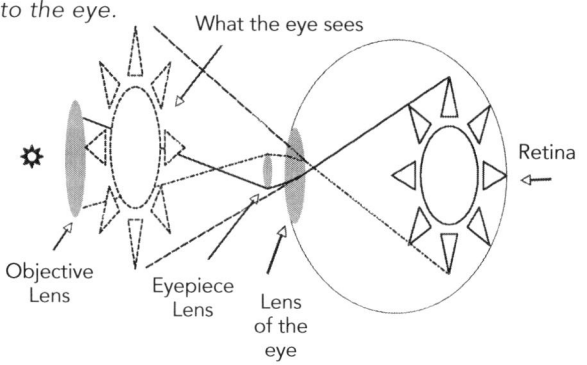

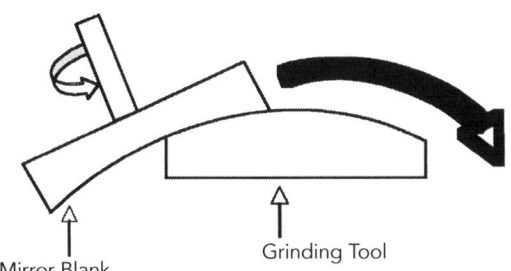

Figure 4. Grinding a mirror by hand (curvature greatly exaggerated).

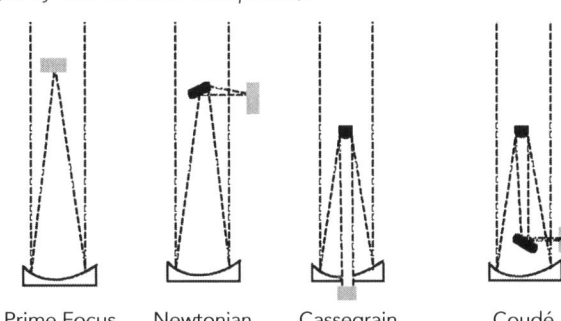

Figure 5. Four designs of reflecting telescopes (Gray bar shows focal plane).

in the world as of 2010 is 40 inches (1.106 meters) at Yerkes Observatory in Wisconsin.

The eye has its own lens, and the telescope has two lenses (or sets of lenses): the objective and the eyepiece. Figure 3 shows how the two-lens telescope delivers a greatly magnified image to the eye. The magnification in diameters is equal to the focal length of the objective divided by the focal length of the eyepiece. For example, the 40-inch telescope at Yerkes has a focal length of 744 inches. With a one-inch eyepiece, this telescope magnifies 744 diameters.

A microscope operates in the same way, except that the object being viewed, instead of distant and dim, is well lit and close to the objective lens.

Diffraction and Refraction

The useful magnification of a telescope is limited by diffraction. Light rays at the edge of the objective lens are diffracted—they are bent around the edge of the lens. These diffracted light rays cause a pattern of light and dark circles around bright images, which will blur adjacent images together. An empirical formula traditionally used to specify the limit of useful magnification is the Dawes Limit (also called the Rayleigh Limit): the resolution in arc-seconds is $4.56/D$, where D is the diameter of the objective in inches; or $11.6/D$, where D is in centimeters. For example, the diameter of the pupil of the human eye when dark-adapted is approximately 8 mm. By the Dawes Limit, the eye can resolve 11.6/.8 (14.5 arc-seconds), or about 1/125 of

Non-Optical Telescopes

Gravitational fields bend light, as predicted by Albert Einstein's general theory of relativity. Hence, large gravitational fields act as lenses. The first test of general relativity was during a solar eclipse in 1919, when the effect of the sun's gravity was to make stars very near the sun's edge appear to be at a small—but measurable— angle further away from the sun than when they are viewed when the sun is not almost in front of them. In effect, the sun acted as a lens and magnified the image of the area around the sun. There are no lenses for radio waves, but radio telescopes that observe radio waves from astronomical objects, such as quasars, do exist. Most radio telescopes use a metal parabolic mirror to reflect the astronomical radio waves to a receiver at the focus of the parabola.

There also exist what might be called sound telescopes. One variety, for picking up sounds from a distance, uses a parabolic dish to reflect sound waves to a microphone at the focus of the parabola. Ultrasound machines, used for monitoring pregnant women, use the woman's own bladder to focus the ultrasound waves onto the receiver.

the diameter of the full moon. The Yerkes telescope can resolve about 0.1 arc-seconds.

A telescope using a lens as its objective is called a *refracting telescope*, since light is "refracted" (bent) by the lens. As of 2010, the 40-inch Yerkes instrument is the largest refracting telescope. A lens that size has to be thick to stand up to gravity, and thick lenses absorb so much light that beyond the size of Yerkes, absorption begins to outweigh the increased light gathered by a wider lens. Hence all current telescopes with objectives greater than 40 inches are "reflecting telescopes" in which the objective is a mirror rather than a lens.

Observer Placement

Unlike a lens, an objective mirror has a parabolic rather than a spherical surface. There is also the mechanical problem of where to place the observer or camera. There are several possibilities, some of which are shown in Figure 5.

One method, called "prime focus," places the photographic film (or other astronomical instrument) inside the path of the incoming light. A few very large reflecting telescopes, such as the 200-inch Hale Telescope at Mount Palomar, actually allow for a human observer to ride in a cage at the prime focus.

A more common arrangement, invented by Isaac Newton and called the "Newtonian," consists of a small flat mirror at an angle, which moves the focal plane to the side of the telescope. Two other common arrangements have a convex mirror at the prime focus, reflecting the light back down the length of the incoming light and also increasing the focal length. In the Cassegrain arrangement, a hole is cut in the middle of the mirror for the light to pass through. In the *coudé* arrangement, the light is reflected one more time into the mounting of the telescope, allowing the use of stationary instruments too heavy to be loaded onto the tube of the telescope.

Further Reading

Alloin, D. M., and Jean-Marie Mariotti. *Diffraction-Limited Imaging With Very Large Telescopes*. Berlin: Springer, 1989.

Edgerton, Samuel. *The Mirror, the Window, and the Telescope: How Renaissance Linear Perspective Changed Our Vision of the Universe*. Ithaca, NY: Cornell University Press, 2009.

Gates, Evalyn. *Einstein's Telescope: The Hunt for Dark Matter and Dark Energy in the Universe*. New York: W. W. Norton, 2009.

Maran, Stephen. *Galileo's New Universe: The Revolution in Our Understanding of the Cosmos*. New York: BenBella Books, 2009.

James Landau

Temperature

Category: Space, Time, and Distance.
Fields of Study: Algebra; Measurement.
Summary: Scientists and mathematicians have developed and investigated a variety of principles and scales associated with the measurement and definition of temperature.

Quantification of temperature is necessary for many reasons, including scientific experiments, weather prediction, and many manufacturing processes. Temperature, by its formal definition, measures the movement of molecules in an object. Greater movement results in higher temperatures; conversely, less movement results in lower temperatures. The byproduct is heat, so temperature is often thought to measure the heat of an object. Mathematicians, many of whom are also physicists, have made significant contributions in quantifying heat and developing the temperature scales widely used in the twenty-first century.

History

Joseph Fourier began heat investigations in the early nineteenth century. His work *On the Propagation of Heat in Solid Bodies* was controversial at the time of its publication in 1807. Joseph Lagrange and Pierre-Simon Laplace argued against Fourier's trigonometric series expansions; however, Fourier series are widely used in a variety of theories and applications in the twenty-first century. Jean-Baptiste Biot, Simone Poisson, and Laplace objected at various times to Fourier's derivation of his heat transfer equations. In 1831, Franz Neumann formulated the notion that molecular heat is the sum of the atomic heats of the components. Studying mixtures of hot and cold water, which did not produce water that was the average of the two temperatures, he concluded that water's specific gravity increases with temperature. This relationship was later shown by other researchers to be true only for a certain range of temperatures. In the late nineteenth century, James Maxwell and Ludwig Boltzmann independently developed what is now known as the "Maxwell–Boltzmann kinetic theory of gases," showing that heat is a function of only molecular movement. Their equations have many applications, including estimating the heat of the sun.

Around the same time, Josef Stefan proposed that the total energy emitted by a hot body was proportional to the fourth power of the temperature, based on empirical observations. In the twentieth and twenty-first centuries, scientists continued to study heat and have developed mathematical and statistical models to estimate heat. These models are used in areas like astronomy, weather prediction, and the global warming debate.

Measuring Tools and Temperature Scales

Heat can be difficult to quantify. Scientists and mathematicians developed many methods and instruments to measure and describe perceived temperature. Some of the earliest were called *thermoscopes*, often attributed to Galileo Galilei. In the early 1700s, Gabriel Fahrenheit created mercury thermometers and marked them with units that became known as "degrees Fahrenheit." He empirically calibrated his thermometer using three values. Icy salt water was assigned temperature zero. Pure ice water was labeled 30. A healthy man would show a reading of 96 degrees Fahrenheit. Later, Fahrenheit would measure the temperature of pure boiling water as 212 degrees Fahrenheit, adjusting the freezing point of water to be 32 degrees Fahrenheit so there was 180 degrees between the freezing and boiling point of water.

Anders Celsius created a different temperature scale in the mid-1700s. The Celsius temperature scale was numerically inverted with respect to Fahrenheit. He used 100 to indicate the freezing point of water and 0 for the boiling point of water. Because there were 100 steps in his temperature scale, he referred to it as a "centigrade" (*centi* means "a hundred" and *grade* means "step"). A few years later, Carolus Linnæus allegedly reversed the scale to make zero the freezing point and 100 the boiling point.

About a century after Celsius created his scale, William Thomson, Lord Kelvin, is given credit for the idea of an absolute zero, a temperature so cold that molecules do not move. The Kelvin scale was precisely defined much later after scientists and mathematicians better understood the concept of conservation of energy. Near-absolute zero conditions produce many interesting problems in mathematics and science. For example, clumping of atoms as they approach an unmoving state can be studied as a classic packing problem, which has extensions in areas like materials science and digital compression. The Kelvin temperature scale uses the same scale as centigrade, with absolute zero about 273 degrees below the freezing point of pure water. Converting from degrees centigrade to Kelvin is as simple as shifting the scale by adding 273.

In the mid-twentieth century, the centigrade scale was replaced with the Celsius scale. The changes were relatively minor, so one estimates the freezing and boiling points of water to be 0 degrees Celsius and 100 degrees Celsius. In actuality, 100 degrees Celsius (the boiling point of water) is now 99.975 degrees Celsius. Converting from degrees Celsius to degrees Fahrenheit, or degrees Fahrenheit to degrees Celsius, involves multiplicative rescaling, not just translation, since 1 degree Celsius is 1.8 times larger than 1 degree Fahrenheit.

Further Reading

Callen, Herbert B. *Thermodynamics and an Introduction to Thermostatistics*. Hoboken, NJ: Wiley, 1985.

Chang, Hasok. *Inventing Temperature: Measurement and Scientific Progress*. New York: Oxford University Press, 2004.

Lipták, Béla G. *Temperature Measurement*. Radnor, PA: Chilton Book Co., 1993.

Pitts, Donald R., and Leighton E. Sissom. *Schaum's Outline of Theory and Problems of Heat Transfer*. New York: McGraw-Hill, 1998.

Quinn, T. J. *Temperature*. San Diego, CA: Academic Press, 1990.

Chad T. Lower

Tides and Waves

Category: Weather, Nature, and Environment.
Fields of Study: Geometry; Number and Operations.
Summary: Mathematicians study and model the forces that cause tides and waves.

Approximately 70% of the Earth's surface is covered with water, most of which is in a constant state of motion. The causes of this motion include the gravitational pull of celestial bodies in space, like the sun and moon; the rotation and shape of the Earth; and the influence of natural phenomena, like wind and earthquakes. Mathematicians have long studied tides and waves, following in the path of ancient scholars and others who sought to understand these phenomena for many spiritual and practical reasons, such as sailing. In the twenty-first century, people still travel both above and below the surface of the oceans for research, commerce, and pleasure, and there are many problems old and new to be explored. Some interesting mathematical investigations related to tides and waves at the start of the twenty-first century include three-dimensional modeling of extreme waves (also called "rogue waves"), such as those observed during the 2004 Indian Ocean tsunami and the Hurricane Katrina storm surges in 2005. Mathematicians, scientists, and engineers have also explored methods and developed technology to harness tide and wave power as an alternative energy source, including methods that actually create waves in addition to using naturally-occurring ones. Some colleges and universities teach courses on tides and waves that involve substantial mathematics. The theme of Mathematics Awareness Month in 2001 was "Mathematics and the Ocean," underscoring the importance and relationship of ocean phenomena and mathematics, as well as the depth and breadth of the topics studied.

Tides

Water in Earth's oceans moves in a variety of ways, including many scales of currents, tides, and waves. Mathematicians and scholars from ancient times up through the Renaissance observed, identified, and quantified tidal patterns. The term "tides" generally refers to the overall cyclic rising and falling of ocean levels with respect to land—though tides have been observed in large lakes, the atmosphere, and Earth's crust, resulting largely from the same forces that produce ocean tides.

The daily tide cycles are caused by the moon's gravity, which makes the oceans bulge in the direction of the moon. A corresponding rise occurs on the opposite side of the Earth at the same time, because the moon is also pulling on the Earth itself. Most regions on Earth have two high tides and two low tides every day, known as "semidiurnal tides," which result from the daily rotation of the Earth relative to the moon. Since the angle of the moon's orbital plane also affects gravitational pull on Earth's curved surface, some regions have only one cycle of high and low, known as "diurnal tides." The height of tides varies according to many variables, including coastline shape; water depth ("bathymetry"); latitude; and the position of the sun, which also exerts gravitational force. "Spring" tides, not named for the season, are extremely high and low tides that occur during full and new moons when the sun and moon are in a straight line with the Earth, and their gravitational effects are additive. A proxigean spring tide occurs roughly once every 1.5 years when the moon is at its proxigee (closest distance to Earth) and positioned between the sun and the Earth. Neap tides minimize the difference between high and low tides. They occur during the moon's quarter phases when the sun's gravitational pull is acting at right angles to the moon's pull with respect to the Earth.

A few of the many contributors to the theory and mathematical description of tides include Galileo Galilei, René Descartes, Johannes Kepler, Daniel Bernoulli, Leonhard Euler, Pierre Laplace, George Darwin, and Horace Lamb. Some mathematicians, like Colin Maclaurin and George Airy, won scientific prizes for their research. Work by mathematician William Thomson (Lord Kelvin) on harmonic analysis of tides led to the construction of tide-predicting machines.

Waves

There are many mathematical approaches to the study of waves in the twenty-first century, and some mathematicians center their research around this topic. In contrast to tides, a wave is a more localized disturbance of water in the form of a propagating ridge or swell that occurs on the surface of a body of water. Despite the fact that surface waves appear to be moving when observed, they do not move water particles horizon-

Officers of the National Oceanic & Atmospheric Administration Corps photographed the devastation caused in New Orleans by the 2005 Hurricane Katrina's storm surges. (NOAA Aviation Weather Center)

tally along the entire path of the wave. Rather, they combine limited longitudinal or horizontal motions with transverse or vertical motions. Water particles in a wave oscillate in localized, circular patterns as the energy propagates through the liquid, with a radius that decreases as the water depth or distance from the crest of the wave increases. Wind is a primary cause of surface waves, because of frictional drag between air and water particles. Larger waves, like tsunamis, result from underwater Earth movements, such as earthquakes and landslides.

The Navier–Stokes equations, named for Claude-Louis Navier and George Stokes, are partial differential equations that describe fluid motion and are widely used in the study of tides and waves. Solutions to these equations are often found and verified using numerical methods. The Coriolis–Stokes force, named for George Stokes and Gustave Coriolis, mathematically describes force in a rotating fluid, such as the small rotations in surface waves. A few examples of individuals with diverse approaches who have won prizes in this area include Joseph Keller, who has researched many forms and properties of waves, including geometrical diffraction and propagation; Michael Lighthill and Thomas Benjamin, who jointly posed the Benjamin–Lighthill conjecture regarding nonlinear steady water waves, which continues to spur research in both theoretical and applied mathematics; and Sijue Wu, who has researched the well-posedness of the fully two- and three-dimensional nonlinear wave problem in various function spaces, using techniques like harmonic analysis. In other theoretical and applied areas, some techniques from dynamical systems theory, statistical analysis, and data assimilation, which combines data and partial differential equations, have been useful for formulating and solving wave problems.

Further Reading

Cartwright, David. *Tides: A Scientific History.* Cambridge, England: Cambridge University Press, 2001.

Johnson, R. I. *A Modern Introduction to the Mathematical Theory of Water Waves.* Cambridge, England: Cambridge University Press, 1997.

Joint Policy Board for Mathematics. "Mathematics Awareness Month April 2001: Mathematics and the Ocean." http://mathaware.org/mam/01.

Sarah J. Greenwald
Jill E. Thomley

Transplantation

Category: Medicine and Health.
Fields of Study: Algebra; Data Analysis and Probability; Number and Operations.
Summary: Locating and allocating available compatible organs is an important task of surgery, as is determining the likelihood of success and survival.

Organ transplantation involves replacing a damaged organ or body part with an organ taken from another body, a location on the patient's own body, or sometimes another source. Relatively common organ transplants include hearts, lungs, livers, corneas, bone marrow, and skin. In the twenty-first century, there are increasing instances of transplantations involving parts that have proven more difficult in the past, including a human face in 2010. Transplantation is one of few medical fields where practice is driven by statistical analysis of large-scale national datasets. Collecting comprehensive data about transplantation in the United States is mandatory, and researchers use statistics to inform clinical practice and national policy. Still, there are too few living and deceased organ donors to meet the need. Optimization tools make the best use of scarce resources, like donated organs. With kidney paired donation, optimization can even increase the supply of available organs. An artificial pancreas employing control theory was under development in 2010.

Statistics

Statistical analyses inform transplant policy and individual decisions. The transplant community seeks equity in allocating organs, so the allocation system is frequently analyzed for gender and racial disparities. Understanding outcomes with and without transplantation helps patients decide if they will benefit from a particular transplant.

Survival analysis is the branch of statistics concerned with the distribution of time to an event. Survival analysis is commonly used in medicine to study time-to-death but can also be used to study time to any event, such as time from joining a transplant waitlist until receiving a transplant. The survival function $S(t) = \Pr(T > t)$ indicates the probability that the random time of an event T is later than a given time t.

Complications

Survival analysis is complicated by censoring; not all patients in a study have reached the event of interest. In a time-to-death analysis, some patients are likely still alive. The first technique for estimating a survival function with censoring was the product-limit estimator of statisticians Edward Kaplan and Paul Meier.

Confounding is another challenge. One could perform a survival analysis of the association between gender and time-to-transplantation to see whether men and women receive transplants at the same rate. However, not all patients are expected to wait the same amount of time. Other factors (such as age and blood type) confound studies of the effect of the factor of interest (gender) on time-to-transplantation. Cox proportional hazards analysis methods, named for statistician David Cox, can account for confounding, using a regression model based on the hazard function $\lambda(t)dt = \Pr(t \leq T < t + dt \mid T \geq t)$, which indicates the instantaneous probability of an event at some time (t) conditional on having survived to at least that time.

Optimization

Donated organs are scarce and each organ must be allocated to one of many potential recipients. Optimization techniques allocate scarce resources by maximizing an objective function. A person's Lung Allocation Score is largest when the transplant has the largest lifespan benefit, and available lungs are offered to the nearby person with the largest score.

Kidney paired donation in which two living donors who are incompatible with their intended recipients exchange kidneys for compatible transplants requires more complex optimization techniques. More people can obtain better transplants when the paired donations are arranged using either a maximum weight matching in a graph or a maximum weight cycle decomposition (if more than two donors and recipients are involved in each exchange). By optimizing an individual's outcome rather than the overall good, a Markov decision process model, named for mathematician Andrei Markov, can determine whether it is better for a patient to accept a certain organ offered or wait until a possibly better organ is offered later. Another Markov decision process model can establish the best time for a patient to receive a liver transplant from a living donor.

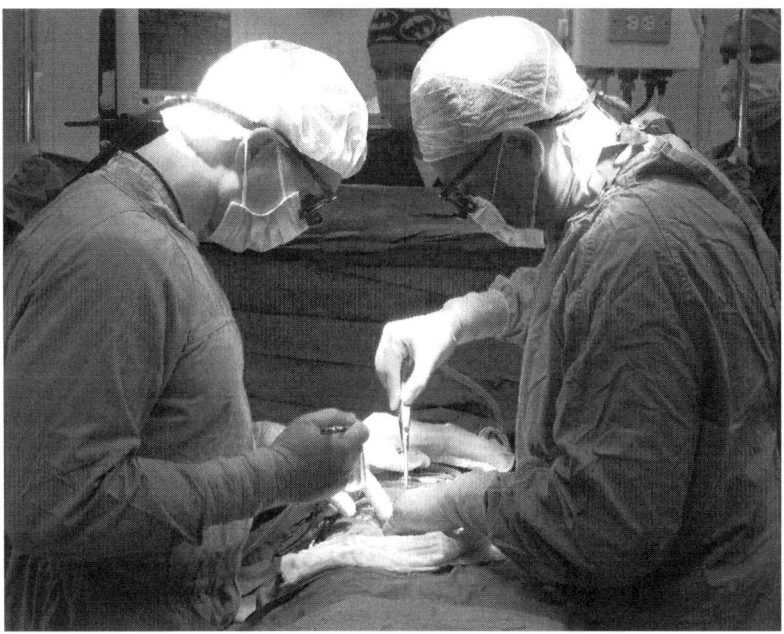

Doctors from Walter Reed Army Medical Center Organ Transplant Service perform Guyana's first kidney transplant operation in 2008. (U.S. Army, Laura Owen)

Control

Control theory studies systems where adjustments over time maintain some desired set point, like a thermostat heating or cooling a room to maintain a comfortable temperature. In transplantation, control theory is used in an experimental artificial pancreas. A healthy person's pancreas maintains blood glucose levels over time by regulating insulin in response to eating a meal or exercising. An artificial pancreas uses a blood glucose monitor and a mathematical control system to drive an insulin pump. The control algorithms are tested on mathematical models of blood glucose levels before being tested in human subjects.

Further Reading

Cox, David R. "Regression Models and Life Tables." *Journal of the Royal Statistical Society* 34, no. 2 (1972).

Harvey, R. A., et al. "Quest for the Artificial Pancreas." *IEEE Engineering in Medicine and Biology Magazine* (March/April 2010).

Kaplan, Edward L., and Paul Meier. "Nonparametric Estimation From Incomplete Observations." *Journal of the American Statistical Association* 53 (1958).

Segev, D. L., S. E. Gentry, D. Warren, B. Reeb, and R. A. Montgomery. "Kidney Paired Donation: Optimizing the Use of Live Donor Organs." *Journal of the American Medical Association* 293 (2005).

Sommer Gentry
Dorry Segev

Ultrasound

Category: Medicine and Health.
Fields of Study: Algebra; Geometry; Representations.
Summary: Ultrasound uses mathematical principles to create images of the human body.

Although ultrasound cannot be heard by humans, it has been produced and used for a vast number of applications in many different fields. In industry, ultrasound has been used as a technique to assess the structural integrity of materials. The interaction between ultrasound and live systems has been studied since the 1920s. During the 1960s, it was used in medicine, initially as a therapeutic option and then later as a diagnostic resource. In the twenty-first century, ultrasound is a major medical imaging technology widely used in clinical facilities around the world because it causes no harm to the human body and results can be achieved in real time, besides the fact it is considerably cheap and easy to use. The available technologies using ultrasound are in constant development. Every new application depends on the advance of computer sciences that work with many concepts of physics and the solution of mathematical problems in this field seems inexhaustible.

Sound is a form of energy consisting of the vibration of molecules of an environment that can be air, water, solid, or biological tissues (such as bones and muscles). This kind of energy propagates across the medium in the form of waves. Sound is a mechanical wave whose fundamental characteristics are amplitude, which is the distance between the highest and lowest point of the wave and frequency, which is the number of cycles that occur in a second, measured in hertz (Hz). Humans are able to detect sounds with a frequency of 20–20,000 Hz–the normal limits of the human hearing. The term "infrasound" refers to sound waves that have a frequency lower as 20 Hz, and sounds with a frequency higher than 20,000 Hz are called "ultrasound." Unlike humans, some animals, such as bats, dolphins, whales, dogs, cats, and mice can hear ultrasound.

Imaging the Human Body

While traversing a material, the properties of ultrasound change in intensity and speed of propagation, which means that ultrasound waves travel at different speeds depending on the material. Consider two samples of human bone, one from a 30-year-old person and the other from an 80-year-old person. If ultrasound waves cross these two bony samples, the speed at which the sound propagates in the bones can be represented algebraically by the following equation:

$$v = \sqrt{\frac{E}{\rho}}$$

where v is the speed of ultrasound in the bone sample, E is the modulus of elasticity of the bone sample, and ρ is the density of the bone sample.

The speed of sound (v) can be calculated by measuring the time required for the wave to propagate through the bone and then dividing by the width of the bone.

Knowing the density of the bones (ρ), this equation could be used to determine the values of the modulus of elasticity (E) that indicates the elastic properties of the bone. In a 30-year-old person, the speed of the sound through the bone is approximately 4000 m/s. In an 80-year-old person, this rate drops to 3800 m/s. This fact means that the higher the speed of the sound through the bone, the better is the quality of bone. A low speed could reveal a bony fragility and a fracture probability. This principle is used in ultrasonometry, a technique used to estimate the bony fracture or osteoporosis risk in patients. Ultrasound medical imaging is one of the most powerful diagnostic tools in modern medicine. Along with other imaging methods, it is based on advanced mathematical techniques and numerical algorithms that are necessary to analyze the data and produce readable pictures or three-dimensional images of inner body structures without surgery or use of radiation. It has been widely used to identify the sex or to detect malformations in fetuses during gestation.

Further Reading

Ammari, Habib. *An Introduction to Mathematics of Emerging Biomedical Imaging*. Berlin: Springer, 2009.

Gibbs, Vivien, et al. *Ultrasound: Physics and Technology*. Philadelphia: Churchill Livingstone, 2009.

Maria Elizete Kunkel

Universal Constants

Category: Space, Time, and Distance.
Fields of Study: Number and Operations; Measurement.
Summary: Universal constants help describe the universe and are believed to be fixed for all times and places in the universe.

A universal constant is a physical quantity whose value remains fixed throughout the universe for all time. However, most constants are known only approximately; humans started measuring them relatively recently and it is an assumption that they are—and have always been—fixed. There may be other assumptions that scientists and mathematicians have implicitly made that turn out to be false and undermine the universality of these constants. For example, the ratio of the circumference of a circle to its diameter in Euclidean space is π, but with Albert Einstein's conceptualization that the universe could have non-Euclidean geometry, this circumference-to-diameter ratio in the real world may be some value not equal π.

The international Committee on Data for Science and Technology defines and modifies physical constants and quantifies their levels of certainty. Three constants in particular are fundamental to the current understanding of the physical world. Together, they underlie the mathematics of gravity, relativity, and quantum physics. They are G (the gravitational constant), c_0 (the velocity of electromagnetic radiation in a vacuum (in other words, the speed of light), and h (Planck's constant).

Universal Constant: "G"

G first appeared in Isaac Newton's famous equation $F = Gm_1m_2/r^2$, which quantifies the force (F) of gravitation between two masses (m_1 and m_2), where r is the distance between their centers of mass. G is approximately 6.67×10^{-11} m^{3}kg^{-1}s^{-2} (meters-cubed per kilogram per second-squared), which is a very small number. Gravity is thus a very weak force. Although every mass is attracted to every other mass, the effects of gravity are obvious only when the masses involved are very large (such as with planets).

Using another of Newton's equations, $F = ma$, it follows that the acceleration due to gravity on Earth is the same for all masses. This acceleration is known as g and its value is around 9.81 ms^{-2} at sea level. This value varies with distance from the Earth's center of mass (r in the equation above), so acceleration due to gravity decreases to around 9.78 ms^{-2} at the top of Mount Everest. Knowing g to be about 9.81 ms^{-2} and the radius of the Earth to be roughly 6,378,000 meters, one can use G to show that the mass of the Earth is about 5.98×10^{24} kg. One can also estimate the mass of the Sun and other celestial bodies, such is the applicability of G.

Universal Constant: "c_0"

The velocity of light in a vacuum, c_0, is probably the most widely known universal constant. Since the length of a meter is defined by it, c_0 is fixed at exactly 299,792,458 ms^{-1}. The constancy (or invariance) of c_0 is a principle that was made famous by Albert Einstein in his theory of special relativity. Einstein's principle states that no matter how fast you or the light source are travelling, you will always measure c_0 to be 299,792,458 ms^{-1}. This principle is counterintuitive, but both the constancy of c_0 and related predictions of relativity theory have been verified empirically. From relativity theory, it is known that as velocity increases, measurements of time and space change because duration and displacement are relative—they depend on how fast one is moving. The amounts by which they change are determined by c_0.

What is actually traveling at c_0 in electromagnetic radiation are massless particles called "photons."

As carriers of the electromagnetic force, all light, electricity, and magnetism are the result of photon motion. The relationship between the photon energies and the frequency of their electromagnetic radiation is the basis of quantum physics and the third constant, h.

Universal Constant: "h"

Named after Max Planck, h has an approximate value of 6.63×10^{-34} kgm^{2}s^{-1}. The units of h can be understood as joule-seconds, also known as "action." This unit is distinct from *power*, which is joules *per* second; for example, 10 joules expended every second for 10 seconds is 100 joule-seconds.

The first appearance of h was in the Planck's relation $E = h\nu$. Planck discovered that photons only had certain discrete energy values, the $E = h\nu$ equation relates the energy (E) of the photon to the frequency (ν) of its electromagnetic radiation. The fact that h exists implies that energy comes in discrete lumps, not in a continuous stream. The unit of h appears in a number of important and fundamental relations, such as Werner Heisenberg's uncertainty principle and Niels Bohr's model of the atom.

Further Reading

Carnap, Rudolf. *An Introduction to the Philosophy of Physics*. New York: Dover Publications, 1995.

Feynman, Richard. *Six Easy Pieces: Essentials of Physics Explained by Its Most Brilliant Teacher*. Jackson, TN: Perseus Books, 1995.

———. *The Character of Physical Law*. Cambridge, MA: MIT Press, 2001.

Finch, Steven. *Mathematical Constants*. Cambridge, England: Cambridge University Press, 2003.

Fritzsch Harald. *The Fundamental Constants: A Mystery of Physics*. Translated by Gregory Stodolsky. Singapore: World Scientific Publishing, 2009.

Magueijo, Joao. *Faster Than the Speed of Light: The Story of a Scientific Speculation.* Jackson, TN: Perseus Books, 2002.

EOIN O'CONNELL

Viruses

Category: Medicine and Health.
Fields of Study: Algebra; Geometry.
Summary: The spread of viruses in a population—and the internal structure of viruses themselves—can be analyzed mathematically to help epidemiologists study viral infections.

A virus is a parasite. It cannot reproduce on its own. Instead, it must invade a cell of another organism and use the host cell's machinery to make copies of itself. The newly replicated viruses then leave the host cell and infect other cells. In the process, the virus often damages the host. For example, different viruses cause measles, polio, and influenza in people; hoof-and-mouth disease in cattle; and leaf curl in many vegetables. Mathematics provides a language to describe viral structures. Furthermore, mathematical models of the spread of a virus in a population are powerful tools in public health policy.

Capsid Geometry

A virus consists of genetic material (either DNA or RNA) surrounded by a protein coat called a "capsid." Viruses have much less genetic material and are much smaller than single-celled organisms like bacteria. With limited genetic material, a virus can encode only a few proteins of its own, and so must use them efficiently. Often, the entire capsid is assembled from many copies of a single protein, which means the capsid should be highly symmetric.

One of the first virus structures to be determined was that of the Tobacco Mosaic Virus (TMV). Copies of the TMV capsid protein are arranged in a helix around the viral RNA. Many other viruses have helical capsids as well. In contrast, poliovirus, the Hepatitis B virus, tomato bushy stunt virus, and other viruses have icosahedral capsids. Figure 1 shows a computer-generated image of the poliovirus capsid with protein subunits colored to highlight the icosahedral symmetry. Other, more complicated capsid shapes are possible.

While the capsids do not have flat triangular faces, they have axes of five-fold rotational symmetry, like those through the vertices of the icosahedron; axes of three-fold rotational symmetry, like those through the centers of the triangular faces of the icosahedron; and axes of two-fold rotational symmetry, like those through the centers of the edges of the icosahedron.

Modeling the Spread of Viruses

Models of virus transmission in a population help researchers understand which interventions might slow the spread of a virus. The SIR model, first proposed by W. O. Kermack and A. G. McKendrick in 1927, is one of the simplest and is suitable for viruses such as measles and influenza. Each person in a population is in one of three categories: (1) susceptible to the virus, (2) infected and infectious, or (3) recovered and immune.

Let S, I, and R be the proportion of the population that is susceptible, infected, and recovered, respectively. The SIR model is given by the following system of differential equations:

$$\frac{dS}{dt} = -\beta SI, \quad \frac{dI}{dt} = \beta SI - \gamma I, \text{ and } \frac{dR}{dt} = \gamma I$$

where the constant β depends on the probability that an infected person transmits the virus to a susceptible person, and the constant γ depends on how long it takes an infected person to recover. This model does not lead to simple expressions for S, I, and R as functions of time but it can be explored computationally. One simple way to do so is to treat time discretely and approximate

$$\frac{dS}{dt}$$

by $(S_{t+1} - S_t)$, where S_t is the value of S at time step t. This method yields the difference equations

$$S_{t+1} = S_t - \beta S_t I_t$$

$$I_{t+1} = I_t + \beta S_t I_t - \gamma I_t$$

and $R_{t+1} = R_t + \gamma I_t$.

The basic SIR model can be modified to fit other scenarios. For example, immunity might wear off over time, or some part of the population might be at higher risk of infection, or a vaccination campaign might begin.

The SIR model assumes that all possible contacts between infected people and susceptible people are equally likely (hence the factor of *SI* in *dS/dt*). Modifying the model to reflect the social structure of the population allows researchers to ask crucial questions. If the supply of influenza vaccine is limited, is it more effective to vaccinate school children, who spread the disease, or the elderly, who may suffer more complications from infection? Will closing airports slow an epidemic enough to justify the costs to travelers? In such situations, mathematical models allow public health officials to test the effects of different interventions before choosing a course of action.

Further Reading

Carrillo-Tripp, Mauricio, et al. "VIPERdb2: An Enhanced and Web API Enabled Relational Database for Structural Virology." *Nucleic Acids Research* 37 (2009).

Keeling, Matt J., and Pejman Rohani. *Modeling Infectious Diseases in Humans and Animals.* Princeton, NJ: Princeton University Press, 2008.

Levine, Arnold J. *Viruses.* New York: W. H. Freeman and Company, 1992.

Catherine Stenson

Vision Correction

Category: Medicine and Health.
Fields of Study: Geometry; Measurement; Problem Solving; Representations.
Summary: Modern optometry depends on precise measurements to construct corrective lenses.

Human vision is subject to a variety of ailments and disorders. Some are congenital; others are age-related. Faulty vision results in blurriness, coupled with headaches and ocular tiredness. However, for many years, humans have been perfecting the art of using external implements to aid vision. Technologies exist in the twenty-first century that can restore perfect vision to people suffering from common vision-related problems, such as myopia or astigmatism. The methods used to diagnose vision issues and to construct corrective lenses rely on precise mathematical measurements and understanding of the geometric principles behind light refraction. Vision may also be modeled in various ways, including using a concept called "orthonormal polynomials," such as the Fourier series and optic wavefronts. This has many applications, including laser vision correction. In stereoscopic vision, two-dimensional projections of the world onto the retina of each eye are combined and compared to form a three-dimensional image. It was once thought of as virtually impossible to cure stereoblindness, but in the early twenty-first century, vision therapists use a variety of techniques to help patients perceive stereoscopic depth in three spatial dimensions.

Lens Power

The optical power of a lens, also known as "dioptic power," "refractive power," or "focusing power," is a measure of the curvature of the lens and the degree to which a lens converges or diverges light. It is equal to the reciprocal of the focal length of the lens in meters. Its unit is "diopter." Prescriptions for eyeglasses specify the optical power of the lenses. The human eye has a refractive power of 60 diopters. Stacking lenses helps to combine their optical power.

Eyeglasses and Bifocals

A simple pair of eyeglasses contains nothing more than two pieces of glass shaped in such a way that they act like a pair of lenses. Lenses exploit the physical property of light called "refraction." Refraction occurs when light travels between mediums of different densities, such as air and glass. The change in the medium causes light to bend in a certain calculable way. This property of lenses is suitable to refocus the image back onto the retina in people suffering from long-sightedness and short-sightedness.

The focal length of a lens in air can be calculated using the lensmaker's equation, given by

$$\frac{1}{f} = (n-1)\left[\frac{1}{R_1} - \frac{1}{R_2} + \frac{(n-1)d}{nR_1R_2}\right]$$

where f is the focal length of the lens, n is the refractive index of the material, R_1 is the radius of curvature of the lens surface closest to the light source, R_2 is the radius of curvature of the lens surface farthest from the light source, and d is the thickness of the lens.

To address people suffering from vision problems such as myopia, hyperopia, and astigmatism, bifocal lenses were invented. These lenses have a section of magnification at the lower portion of the frames to allow the wearer to read small print. Benjamin Franklin is generally associated with the invention of the first pair of bifocals.

Contact Lenses

Contact lenses are corrective or cosmetic lenses placed on the cornea of the eye. Their performance is similar to that of eyeglasses but they can be shaped somewhat differently. Spherical lenses are the typical shape of contact lenses on both the inside and the outside surfaces, whereas toric contact lenses, often used for people with astigmatism, are created with curvatures at different angles and cannot move on the eye. Contact lenses are extremely lightweight and are virtually invisible when compared to eyeglasses. However, they are also not held in place by a rigid framework like glasses. Mathematical models are useful for understanding the various movements of lenses within the eye, especially hard contact lenses.

In the twenty-first century, technology has advanced to a level where it is possible to imprint electronics onto the contact lenses themselves, resulting in the ability to project a virtual display onto the eye directly. While this technology by itself does not directly correct any vision problems, it could be used to assist people in their everyday activities, such as locating objects, or reading street signs by magnifying letters.

LASIK

Laser-assisted in situ keratomileusis (LASIK) is becoming an increasingly popular alternative to contact lenses and eyeglasses. LASIK is a type of refractive surgery performed using a laser. A "laser" (Light Amplification by Stimulated Emission of Radiation) is a highly concentrated beam of light capable of focusing high energy in a small area.

The technology was invented by a Colombia-based Spanish ophthalmologist Jose Barraquer. His technique involved cutting thin flaps in the cornea and altering its shape. After the laser was invented, Dr. Bhaumik, in

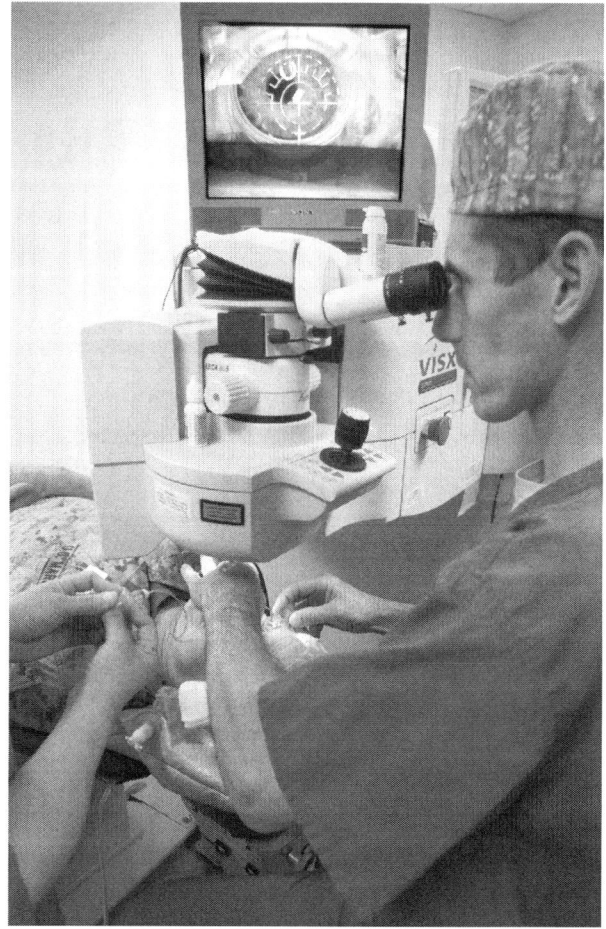

LASIK VISX surgery being performed by a U.S. Navy surgeon. A close-up of the eye is seen on the monitor. (U.S. Navy, Brien Aho)

1973, announced the breakthrough in using lasers to treat vision problems.

LASIK involves creating a flap of corneal tissue, remodeling the cornea underneath the flap with the help of a laser, and then repositioning the flap. Mathematical computations are used to determine the depth of the cuts used in the surgery, and these are often a function of the average cornea thickness of 550 micrometers. One alternative is to leave some fixed tissue depth.

Further Reading

Barry, Susan. *Fixing My Gaze: A Scientist's Journey Into Seeing in Three Dimensions.* New York: Basic Books, 2009.

Dai, Guang-ming. "Wavefront Optics for Vision Correction." *SPIE Press Monograph* PM 179 (2008).

Hecht, Eugene. *Optics*. 4th ed. Addison Wesley, 2002.

Ashwin Mudigonda

Volcanoes

Category: Weather, Nature, and Environment.
Fields of Study: Data Analysis and Probability; Geometry.
Summary: Mathematical models and data analysis can help geologists better understand the activity of volcanoes and the fluid dynamics of their eruptions.

Volcanoes are openings of channels connecting the molten interior of a planet with its surface. Active volcanoes emit magma, ash, and gasses, and inactive volcanoes are reminders of past eruptions, consisting of solidified lava and ash. The science of studying volcanoes is known as "volcanology." Many scientists and philosophers throughout history, including mathematicians Johannes Kepler and René Descartes, theorized about their nature and formation. Mathematics continues to play a role in modern volcanology through both the coursework and degrees that are required and in the mathematical research prevalent in the exploration of various volcanic phenomena. Computer-based numerical simulations and digital imagery, often from satellite observation, combined with mathematical and statistical methods, such as neural networks and data mining, are increasingly used to model, describe, and visualize the complex mathematical representations of volcanic processes. Predicting eruptions is also a challenge, which is necessary not only for safety and response at the time of the eruption but also for larger issues such as global climate change. Benjamin Santer of Lawrence Livermore National Laboratory, who specializes in mathematical and statistical analyses of climate data, has used volcanoes as one variable in explaining climate change. Scientists at the Yellowstone Volcano Observatory also collect data to monitor and mathematically study the enormous Yellowstone caldera, sometimes known as the Yellowstone supervolcano.

Measuring Volcanoes

The most destructive volcanic effect comes from pyroclastic flow, which is a mixture of solid to semi-solid fragments of rock, ash, and hot gases that flows down the sides of the volcano. It is a type of gravity current, similar to an avalanche, that can be modeled with theories and equations from fluid dynamics. A useful metric for comparing eruptions is the volume of volcanic ejecta. For example, the 1980 eruption of Mount St. Helens produced about 1.3 cubic kilometers of ash, but the ancient eruption of the Toba volcano on Sumatra around 75,000 years ago produced more than two thousand times more ash. It is possible to measure the fragmentation of the airborne volcanic matter, called "tephra," even for ancient eruptions. Fragmentation is associated with the strength of the volcanic explosion. The dispersion of tephra over an area has been found to be related to the height of the eruption column. Finding and analyzing dispersion allows estimation of heights for ancient eruptions and an additional way to mea-

After the 1980 eruption of Mount St. Helens, 24 square miles were filled by a debris avalanche. (U.S. Geological Services)

sure heights for modern eruptions. Volcanologists have created the Volcanic Explosivity Index (VEI), which takes into account the volume of ash and the height and duration of the eruption. There are nine types of volcanoes according to VEI, scaled 0–8. For example, the low-strength, low-height Type 0 is called "Hawaiian," and the high-strength, low-fragmentation Type 6 through Type 8 are called "Plinian eruptions," named for Roman historian Pliny the Younger, who described in detail the first century eruption of Mount Vesuvius that destroyed Pompeii. Plinian eruptions can have global environmental effects. Similar to the Richter scale, VEI is logarithmic: each level type is about 10 times greater in magnitude than the previous level.

Geometry of Volcanoes

Shapes of volcanoes depend on their explositivy, viscosity of magma, the composition of the surrounding crust, and other geological factors. The familiar, iconic cone shape such as Mount Fuji defines a "stratovolcano," so named because of its many layers (or "strata") of ash and hardened lava. Eruptions of these volcanoes have high explosivity and low-viscosity lava, making lava and tephra deposit near the opening in layers of diminishing thickness, thus forming the cone.

In contrast, broad, very fluid lava fields produce shield volcanoes that resemble a rather flat warrior shield. Lava domes, as the name suggests, are proportionally higher than shield volcanoes and more rounded than cone volcanoes, resembling semispheres. Lava domes are formed by high viscosity lava combined with low explosivity, where lava either accumulates under the crust and pushes it up, or flows over the crust and solidifies in the dome shape.

Eruption Forecast

Because volcanic eruptions depend on many variables, eruption forecasting relates to such areas of science and mathematics as chaos theory and systems science. Overall, prediction means collecting multi-variate data in volcano observatories and matching variable patterns to those that occurred before eruptions of similar types of volcanoes in the past. For example, the pattern of earthquakes becoming stronger and shallower with time, called "earthquake swarm," can be used to forecast the eruption time. Mathematical models of volcanoes are based on equations from thermodynamics, fluid dynamics, and solid mechanics. The systems science principles of prediction describe qualitative trends in variables. For example, the principle of coinciding change says that unrelated, co-evolving trends in several parameters are more significant than changes in any one parameter.

Further Reading

Marti, Joan, and Gerald Ernst. *Volcanoes and the Environment*. Cambridge, England: Cambridge University Press, 2005.

Zeilinga de Boer, Jelle, and Donalt Sanders. *Volcanoes in Human History*. Princeton, NJ: Princeton University Press, 2002.

Maria Droujkova

Water Quality

Category: Weather, Nature, and Environment.
Fields of Study: Algebra; Data Analysis and Probability; Measurement.
Summary: Water quality standards and data are mathematically modeled and analyzed to help keep drinking water safe.

Water is fundamental for human life. Approximately 70% of the Earth's surface is covered by water but only a very small fraction is consumable fresh water, and much of that has chemical or biological contaminants. Drinking water comes from a variety of sources. Underground water, such as aquifers or springs, may be tapped by wells; surface water, such as rivers and streams, are diverted for use; precipitation may be collected or allowed to flow into other sources; and plants may be processed for moisture. Desalinization (the process of removing salt from water) makes seawater drinkable. Waterborne diseases in open water sources like rivers are endemic to many parts of the developing world. Natural disasters may spread contamination via flooding. Some global warming researchers predict that increased rainfall, flooding, and warmer weather will result in more waterborne disease worldwide. In developed countries, water is commonly piped to end users and may be recycled via sewage treatment. The standards for potable water in many countries are set by government agencies, though the regulation of bottled water differs from piped and well water. Even in nations with extensive closed water distribution systems and sewage treatment, contamination occurs in a number of ways, including agricultural runoff, dumping of manufacturing byproducts into streams and rivers, and degradation of systems that may contain outdated materials such as lead. One of the Millennium Development Goals adopted by the United Nations and other international organizations is to cut in half the proportion of people that do not have reliable access to safe drinking water by 2015. Mathematicians and mathematical methods contribute significantly to the discovery, testing, and delivery of potable water.

How Safe is Your Drinking Water?

The Environmental Protection Agency (EPA) sets the standards for drinking water in the United States. For each potentially harmful substance, the EPA identifies the maximum contaminant level (MCL) allowed and the maximum contaminant level goal (MCLG). The MCLG is the level below which there is zero expected risk to human health. While it would be best to have levels of a substance like arsenic at or below the MCLG, the EPA sets MLC requirements at concentrations that can be higher. U.S. citizens who receive water from a community water system should receive a Water Quality Report each year. Those curious about water quality at work may request a copy from the building owner. Each report includes the source of the water (such as a river or lake); a list of all detected regulated contaminants and their levels; potential health effects of contaminants detected that violate the standards; information for people with weakened immune systems; and contact information for the company or agency that supplies the water. The report will alert the public to violations of the EPA safe drinking water standards and, equally important, will list information about potentially harmful substances that are below the legal limit. For example, a report may list arsenic, describe that it is measured in parts per billion (ppb), give the highest level measured, and list the range measured in the water. The report will also provide the MCLG (0.0 ppb for Arsenic) and the MCL (10.0 ppb). If the report states that the water ranges from 0.5 to 2 ppb for arsenic, water consumers will know that it is safe to drink according to EPA standards. However, upon comparing the MCL and MCLG, consumers may consider drinking water from other sources or request

Providing access to safe drinking water around the world is one of the goals of the United Nations. (Photos.com)

additional information from the water company since 0.5 ppb is higher than the 0.0 ppb MCLG.

Mathematical Analysis and Modeling

The management of water resources is increasingly reliant on mathematical modeling and analysis. For example, the dynamics and kinetics of surface water, along with distributions and dispersal over time of contaminants, have been extensively modeled and simulated. Reactive transport (RT) models use coupled equations to examine particle transportation through porous surfaces, which are widely used to model infiltration of contaminants into ground water. They may utilize mathematical and statistical concepts such as stochastic differential equations, which can be traced in part to physicist Paul Langevin's work on the mathematical theories of dynamic molecular systems. Animal behavioral responses to variables like water quality have been successfully modeled using the Eulerian–Lagrangian–Agent Method (ELAM). The Eulerian framework, named for mathematician Leonhard Euler, mathematically models environment factors affecting the animal agents, while the Lagrangian framework, named for mathematician Joseph Lagrange, governs the perception and movement of individual agents.

Near-continuous water quality monitoring provides a wealth of data and facilitates time series analyses and other statistical models of water quality as functions of variables like land use and precipitation patterns, as well as other measurable human behaviors and natural occurrences. Model calibration, verification, and sensitivity analysis often require comparing mathematical equations and simulation results with observed data. Mathematicians, engineers, and scientists have improved systems for remote water quality monitoring and assessment using data, mathematical methods, and theories from many sciences. Some applications include remote automated stations with the ability to wirelessly network and transmit data, artificial intelligence algorithms that can adaptively sample in response to problems or concerns, and satellite or aircraft observation and analysis of large areas.

These analyses also influence public policy and legislation, such as the U.S. Safe Drinking Water Act and the Clean Water Act. Scientists in many fields continue to seek methods to provide easily accessible clean water for everyone.

Further Reading

Chapra, Steven. *Surface Water-Quality Modeling*. Long Grove, IL: Waveland Press, 2008

U.S. Environmental Protection Agency (EPA). *Drinking Water and Health: What You Need to Know!* Washington, DC: EPA, 1999. http://www.epa.gov/safewater/dwh/dw-health.pdf.

Christine Klein

Weather Forecasting

Category: Weather, Nature, and Environment.
Fields of Study: Connections; Data Analysis and Probability; Problem Solving; Representations.
Summary: Accurate weather forecasting requires the use of advanced mathematical models and powerful supercomputers to handle the vast number of calculations.

Weather prediction, or forecasting, is the application of science and technology to predict the future state of the atmosphere at a given location using available past and present data from the surrounding area. The word "weather" describes the state of the atmosphere at a particular time, or short time period, while the word "climate" is an average of these conditions over long time periods—often months or years. The weather is typically described in terms of temperature, wind speed, wind direction, air pressure, density, and atmospheric composition (for example, water vapor, liquid water, or carbon dioxide content). The intensity of solar and terrestrially emitted radiations is also a fundamental determining factor. A forecast typically includes the prediction of these meteorological variables and helps people make more informed daily decisions that may be affected by the weather. Moreover, it helps predict dangerous weather phenomena, such as hurricanes, which might endanger human life.

History

People have tried to forecast the weather for thousands of years and throughout history, farmers, hunters, warriors, shepherds, and sailors understood the importance of accurate weather predictions for planning daily activities. Ancient civilizations appealed to the gods of the sky: the Egyptians looked to Ra, the sun god; the Greeks sought out Zeus; and in the ancient Nordic culture, Thor was believed to govern the air with its thunder, lightning, wind, rain, and fair weather. The Aztecs used human sacrifice to satisfy the rain god, Tlaloc, while Native American and Australian aborigines performed rain dances.

The Babylonians were predicting the weather from cloud patterns as well as astrology by 650 B.C.E., but the earliest scientific approach to weather prediction occurred circa 340 B.C.E. when Aristotle described his theories about the earth sciences and weather patterns in *Meteorologica*. The ancient Greeks invented the term "meteorology," which derives from the Greek word *meteoron* which refers to any phenomenon in the sky. The Greek philosopher Theophrastus, one of Aristotle's successors, compiled the ultimate weather text *The Book of Signs*, which contained a collection of weather lore and forecast signs and served as the definitive weather book for over 2000 years.

Weather forecasting advanced little from these ancient times to the Renaissance. Beginning in the fifteenth century, Leonardo da Vinci designed an instrument for measuring humidity, Galileo Galilei invented the thermometer, and his student Evangelista Torricelli came up with the barometer. With these tools, people could objectively monitor the atmosphere. In 1687, Sir

Isaac Newton published the physics and mathematics that govern the motion of all bodies and can be used to accurately describe the atmosphere. To this day, his principles are the foundation for modern mathematical analysis and computer prediction of weather.

However, scientifically accurate weather forecasting was not feasible until the early twentieth century, when meteorologists were able to collect and organize data about current weather conditions from observation stations in a timely fashion. Vilhelm and Jacob Bjerknes developed a weather station network in the 1920s that allowed for the collection of regional weather data. The data collected by the network could be transmitted nearly instantaneously by use of the telegraph, invented in the 1830s by Samuel F. B. Morse. This system allowed knowledge of the weather conditions upwind to be incorporated into downwind forecasts, improving their quality.

Great progress was made in the science of meteorology during the twentieth century. The possibility of numerical weather prediction was proposed by Lewis Fry Richardson in 1922, although computers did not yet exist. It was consequently impossible to perform the vast number of calculations required to produce a forecast before the predicted events actually occurred. Practical use of numerical weather prediction began in 1955, spurred by the development of programmable electronic computers.

Numerical Weather Prediction

Numerical weather prediction is the science of forecasting weather using computer simulations built from mathematical models. In this process, the atmosphere is divided into a three-dimensional lattice of grid points, and at each point the various atmospheric variables of interest are represented. These values are initialized with a state determined through analysis of past and present conditions. This state is then evolved forward into the future by solving, at each grid point, the classical laws of (fluid) mechanics and thermodynamics, which are known to accurately approximate the behavior of the atmosphere. The output from the model provides the basis of the weather forecast.

The equations that govern how the state of a fluid changes with time contain many variables and require a great deal of computer processing resources to solve. Weather prediction centers have access to supercomputers containing thousands of processors on which to

Weather forecasters prepare their forecasts at PC workstations with weather analysis software. (NOAA Aviation Weather Center)

run a forecasting model. The required calculations are shared among the processors and computed simultaneously to produce a complete forecast in a fraction of the time possible with a single computer. This system is essential to ensure that an accurate prediction can be made within a useful time frame.

Good weather forecasts depend upon an accurate knowledge of the current state of the weather system, also called the "starting point" or "initial condition." The initial conditions are determined from global measurements of the state of the atmosphere. Surface weather observations of atmospheric pressure, temperature, wind speed, wind direction, humidity, and precipitation are made near the Earth's surface by trained observers, automatic weather stations, or buoys. The initial state has a degree of uncertainty since there are an insufficient number of measurements to initialize all meteorological variables at every grid point. Furthermore, the locations of the measurements do not usually coincide with the numerical grid points and there is also a degree of error in the actual measurement. The problem of determining the initial conditions for a forecast model is very important, highly complex, and has become a science in itself (known as "data assimilation").

The atmosphere is an incredibly complex dynamical system and the approximation of its behavior is only compounded by the inability to measure its state at each and every grid point in the model. The limit on useful weather forecasts using present technology is typically one week. The forecast errors are initially localized, leading to incorrect predictions in small regions, but are generally accurate enough to be useful in most of the forecast area. The longer the simulation is run, the more the measurement and model approximation errors begin to dominate the calculation. However, steady improvements in computer power and prediction models in the twenty-first century have led to a three-day forecast being as accurate as a two-day forecast from the 1990s. Weather forecasting centers are constantly reviewing the accuracy of their forecasts and set themselves annual targets for accuracy improvements.

The raw output from the simulation is often modified before being presented as a forecast. Modifications include either the use of statistical techniques to remove known biases in the model or adjustments to take into account consensus among other numerical weather predictions. Accurate forecasts of precipitation for a specific location are particularly challenging because of the chance that the rainfall may fall in a slightly different place (such as several kilometers away) or at a slightly different time than the model forecasts, even if the overall quantity of precipitation is correct. Therefore, daily forecasts give fairly precise temperatures but put probabilistic values on quantities such as rain, based on knowledge of the uncertainty factors in the forecast.

Probability of Precipitation

A Probability of Precipitation (PoP) is a formal measure of the likelihood of precipitation that is often published from weather forecasting models, although its definition varies. In U.S. weather forecasting, PoP is the probability that greater than 1/100th of an inch of precipitation will fall in a single spot, averaged over the forecast area. For instance, if there is a 100% probability of rain covering one side of a city and a 0% probability of rain on the other side of the city, the PoP would be 50%. A 50% chance of a rainstorm covering the entire city would also lead to a PoP of 50%. The mathematical definition of PoP is defined as PoP = $C \times A \times 100$, where C is the confidence that precipitation will occur somewhere in the forecast area, and A is the percent of the area that will receive measurable precipitation, if it occurs at all.

For example, a forecaster may be 40% confident that precipitation will occur and that, should rain happen to occur, it will happen over 80% of the area. This results in a PoP of 32%: $0.4 \times 0.8 \times 100 = 32\%$.

The Future

Over the years, the quality of the models and methods for integrating atmospheric observations has improved continuously, resulting in major forecasting improvements. The power of supercomputers has increased dramatically, allowing for the use of much more detailed numerical grids and fewer approximations in the operational atmospheric models. Small-scale physical processes (such as clouds, precipitation, turbulent transfers of heat, moisture, momentum, and radiation) have been more accurately represented within the model. Finally, the use of increasingly accurate methods of data assimilation and the integration of satellite and aircraft observations has resulted in improved initial conditions for the models, which ultimately lead to a better forecast.

Further Reading

Kalnay, Eugenia. *Atmospheric Modeling, Data Assimilation and Predictability*. Cambridge, England: Cambridge University Press, 2003.

Pasini, Antonello. *From Observations to Simulations: A Conceptual Introduction to Weather and Climate Modeling*. Singapore: World Scientific Publishing, 2005.

SILVIA LIVERANI

Weather Scales

Category: Weather, Nature, and Environment.
Fields of Study: Algebra; Connections; Measurement.
Summary: Weather scales and tools are used to help measure and classify atmospheric conditions.

Weather affects virtually every aspect of human life, including afternoon showers that might inconvenience commuters; tremendously destructive episodes, like hurricanes; and long-term occurrences, like drought, which impact agriculture and increase the likelihood of other events like wildfires. Meteorology is an interdisciplinary science that focuses on weather and short-term forecasts, typically up to a few weeks. Climatology is a science that looks at long-term average weather. In fact, many define the word "climate" in terms of the average of weather over time, both locally and globally. Mathematics plays a critical role in weather science, enabling people to quantify, compare, model, and predict weather. Valid and reliable comparisons are facilitated by the development of scales and standard systems of quantification, along with mathematical and statistical models that use those measures.

It is thought that some ancient peoples had methods for predicting the weather, though historical evidence is mixed. In the early twentieth century, mathematician Vilhelm Bjerknes and colleagues examined several measurable variables of weather and derived equations to connect them to one another. Mathematician Lewis Richardson, who contributed significantly to mathematical weather prediction and pioneered the use of finite differences in the field, reformulated the Bjerknes equations. However, they remained impractical for rapid forecasting until the introduction of computers. Another product of his work, the Richardson number, is a function of density and velocity gradients that helps predict fluid turbulence in weather and other applications. Mathematicians continue to contribute and modern forecasting involves a wide variety of mathematical techniques and models, drawing in depth from such areas as chaos theory, data assimilation, statistical analyses, scale cascades of error (related to the so-called butterfly effect), numerical analysis, vectors, fluid dynamics, and entropy. Climatologists, scientists, and mathematicians also research related phenomena like geomagnetic and solar storms.

Temperature, Pressure, and Humidity

One of the most pervasive and intuitively obvious variables used to characterize the weather is air temperature—along with air pressure and humidity in most modern reports and forecasts. Strictly speaking, air temperature is a measure of the average kinetic energy of the air molecules, measured by a variety of types of thermometers. The most common scales used to quantify temperature are the Celsius (or centigrade) scale used throughout most of the world and the Fahrenheit scale used primarily in the United States. Atmospheric pressure is measured by a barometer, whose invention is attributed to various sources including Galileo Galilei and mathematicians Gasparo Berti and Evangelista Torricelli.

There are many common units for pressure, including inches of mercury, pounds per square inch, pascals, named for mathematician Blaise Pacsal, and atmospheres. One atmosphere is defined as the mean atmospheric pressure at mean sea level, originally measured with respect to the latitude of Paris, France. Millibars are often used in weather reports and forecasts. A hygrometer measures the amount of water vapor in the air. How much water vapor the air can hold is a function of temperature and relative humidity expresses the quantity of water vapor as a unitless fraction or percentage of the possible amount of water for a given temperature. Humidity can be used in probability models to predict precipitation, dew, and fog. Further, high humidity changes the subjective feeling of the air temperature for people because high humidity reduces

Table 1: Fujita scale of tornado strength.

Scale	Wind Speed (km/hr)	Damage
F-0	65–118	Light
F-1	119–181	Moderate
F-2	182–253	Considerable
F-3	254–332	Severe
F-4	333–419	Devastating
F-5	420–513	Incredible

Table 2. Saffir-Simpson scale of hurricane strength.

Scale Number	Wind Speed (km/hr)	Storm Surge (meters)	Central Pressure (millibars)	Damage
1	121–154	1–2	≥ 980	Minimal
2	155–178	2–3	965–979	Moderate
3	179–210	3–4	945–964	Extensive
4	211–250	4–6	920–944	Extreme
5	>250	>6	<920	Catastrophic

the evaporation of sweat. This effect is quantified as a heat index, with assumptions about many variables such as wind speed, body mass, clothing, physical activity, and exposure to sunlight. A similar concept is wind chill, which relates the subjective perception of cold. Scientist Robert Steadman has researched and mathematically modeled both of these effects and they have become a common part of weather forecasts.

Wind

Another weather variable is wind speed. In 1805, Sir Francis Beaufort, an Irish hydrographer, developed what is now called the Beaufort scale to describe and categorize the strength of the wind. The scale has 13 points ranging from zero (calm air) to 12 (hurricane-force winds). On the scale, the Beaufort number two is identified as a "light breeze," with wind speed 6–11 kilometers per hour (km/hr) producing wind that is felt on the face, leaves that rustle, movement of a wind vane, and on the water, small, short wavelets that do not break. Further along the scale is Beaufort number five, a "fresh breeze," with wind speeds between 29 and 38 km/hr. At this point, small, leafy trees will sway, moderate waves become longer, and there are many whitecaps and some spray. Wind speeds between 62 and 74 km/hr are classified as a "gale," Beaufort number eight. Twigs and small branches break off trees. At sea, there are moderately high waves of greater length. Beaufort number 10 is used when wind speeds are between 89 and 102 km/hr and are "storm-force" winds. Trees are broken and uprooted and structural damage occurs. At sea, there are very high waves with overhanging crests and visibility is reduced.

The terms and descriptions make it clear that as wind speed rises so does its destructive power. In fact, the force exerted by wind increases as the square of the velocity such that a doubling of the wind's velocity leads to a quadrupling of the force: $F \sim V^2$. Some of the most powerful winds experienced on Earth are found in hurricanes and tornadoes. Their destructive power can be astounding and has been the subject of much study and research. The Fujita scale, presented in Table 1, is used to categorize tornado strength in terms of rotational wind speed (given in km/hr) and damage inflicted by the wind. While tornadoes are generally associated with severe thunderstorms and are seldom more than 1.5 km in diameter, hurricanes can involve whole systems of thunderstorms and may be several hundred kilometers in diameter. The Saffir–Simpson scale, used to categorize hurricanes, is presented in Table 2.

Further Reading

Ahrens, Donald C. *Meteorology Today*. Belmont, CA: Thompson Brooks/Cole, 2007.

Lynch, Peter. *The Emergence of Numerical Weather Prediction: Richardson's Dream*. Cambridge, England: Cambridge University Press, 2006.

Moran, Joseph P., and Lewis W. Morgan. *Meteorology: The Atmosphere and the Science of Weather*. Edina, MN: Burgess Publishing, 1986.

Mark Roddy
Sarah J. Greenwald
Jill E. Thomley

Weightless Flight

Category: Travel and Transportation.
Fields of Study: Algebra; Geometry.
Summary: The forces to experience the sensation of weightlessness, or zero-*G*, can be calculated and achieved in a variety ways.

Gravity is the mutual attraction of two masses. Important aspects of the mathematics and the theory of gravity were described centuries ago by Galileo Galilei and Isaac Newton. Albert Einstein's work was critical to the modern understanding of gravity and weightlessness. Mass is the measure of the amount of matter in an object. For living beings, weight can be thought of as the subjective experience of muscles resisting the pull of the much larger Earth on their smaller masses. On the Earth's surface, gravitational acceleration is about 9.8 meters per second (one gravity or *g*). Other planets have different gravity. For example, an Earth person would feel about 2.5 times heavier on Jupiter. Infants learn to accommodate gravity's pull when performing the activities of daily life until the force feels natural and largely unnoticed. However, sometimes people experience other forces acting on their bodies that counter the pull of gravity and change their perceptions of weight. For example, the quick start or stop of an elevator can make a person feel heavier or lighter. Roller coasters purposely induce similar effects for amusement. Parabolic drops, turns, and loops exert temporary linear or angular forces on a moving body, some of which act along a different directional vector than gravity and combine mathematically to alter the body's perception of weight. Mathematicians, scientists, and engineers precisely calculate the net effect of gravity and other forces on objects for a wide range of applications, such as banked curves on racetracks and highways, the movement of subatomic particles, launching spacecraft to the moon, and of course, ever more thrilling amusement park rides.

Zero-G

The planet's mass exerts a strong gravitational pull even on objects in space. This force is what keeps satellites in position. However, many people have seen video images of astronauts who are floating around as if they are weightless. This effect is known as *zero-G* or, more accurately, "microgravity" (about 1×10^{-6} g). Like roller coasters, this effect results from a combination of forces acting on the body. At any given instant in time, the astronauts are accelerating freely toward the Earth inside an object that is accelerating freely at the same rate. They can be visualized in that instant as falling on a straight line drawn from the spaceship to the Earth, perpendicular to a tangent line drawn at the ship's current position in its curved orbit. However, the ship's directional vector is constantly changing because of its curved orbit, so it perpetually "falls" in a new direction—around the Earth, instead of toward it. The spacecraft's precisely calculated inertial trajectory effectively counters the astronauts' constant "falling." As a result, the astronauts do not move with respect to their immediate surroundings, so they look and feel as if they are floating weightlessly. A spacecraft lands by altering its curved orbit so that the gravity is no longer sufficiently opposed.

Free-fall or zero-G can be achieved in several ways without leaving Earth's atmosphere. NASA's Neutral Buoyancy Simulator uses the world's largest indoor pool, containing over six million gallons of water, to simulate weightlessness without flying or falling, while their Zero Gravity Research Facility can achieve just over five seconds of free fall in a 467-foot long steel vacuum chamber, which is used to test microgravity effects on phenomena such as combustion and fluid physics. As part of a series of experiments in the 1960s, Air Force Captain Joseph Kittinger parachuted from a gondola at an altitude of almost 103,000 feet. He achieved

Astronauts aboard an aircraft that flies a parabolic pattern to provide weightlessness training. (National Aeronautics and Space Administration)

a speed of over 600 miles per hour on his descent but he reported having no real subjective sensation of the incredible speeds. Standard aircraft can be used to create brief periods of weightlessness, about 30 seconds, by flying in a parabolic pattern or "Kepler curve," named for Johannes Kepler. NASA uses this method to train astronauts, and the weightless effects seen in the 1995 movie *Apollo 13* were produced using parabolic flight. Several commercial companies also offer the experience to the general public. A privately funded experimental "spaceplane" called SpaceShipOne achieved suborbital flight in 2004. A revised commercial version called VSS Enterprise flew for the first time in 2010 and is taking reservations for future commercial flights that will launch passengers into suborbital space.

Further Reading

Clement, Giles, and Angeli Bukley. *Artificial Gravity*. New York: Springer, 2007.

Erickson, Lance. *Space Flight: History, Technology, and Operations*. Lanham, MA: Government Institutes, 2010.

Sparrow, Giles. *Spaceflight: The Complete Story From Sputnik to Shuttle—and Beyond*. New York: Dorling Kindersley Publishers, 2007.

JULIAN PALMORE

Wind and Wind Power

Category: Weather, Nature, and Environment.
Fields of Study: Data Analysis and Probability; Geometry; Measurement.
Summary: Wind and wind power have been mathematically studied for centuries as an energy source and promise to be increasingly important energy sources.

Wind is omnipresent. There are few parts of the world that are not affected by the wind, from the pleasant breezes off a lake to the terrifying destruction of hurricanes and tornados. Historically, wind was one of the most important sources of energy; it drove sailing ships and was key to driving some pre-industrial revolution machines, such as windmills. Being able to master the wind was a key component in the fate of empires. For example, in 1588, it is said that the Spanish Armada of Catholic King Philip II was defeated by a "strong Protestant wind" that forced his fleet off course and prevented a vulnerable England under the reign of Queen Elizabeth I from being invaded. In the wake of the steam engine, developed by James Watt in the 1760s, and the emergence of coal-powered machines during the Industrial Revolution, the age of wind and sail began to decline for much of the industrialized world. Many cite this shift to fossil fuel sources as a cause of the rise in carbon dioxide, other greenhouse gasses (GHGs), and the global warming phenomenon, and there is a movement toward returning to wind as one source of clean energy.

Mathematicians and scientists have long been involved in the study of wind and wind energy. Posidonius of Rhodes (c. 135–51 B.C.E.) theorized about clouds, mist, wind, and rain. Francis Beaufort (1774–1857) developed a mathematical scale to describe wind speed. Twenty-first-century engineer Michael Klemen has explored mathematical issues of wind data acquisition as a function of time and estimated wind resource availability for power generation. Mathematicians continue to contribute to these fields and to the exploration of related phenomena like solar winds, which are believed to have first been observed by astronomer John Herschel during his observations of Halley's comet in 1835.

History

Seventeenth-century mathematician Evangelista Torricelli was reputed to be skilled in making instruments and he is often credited with inventing the barometer. He also conducted research about weather and is believed to have given the first correct explanation of wind when he said, "winds are produced by differences of air temperature, and hence density, between two regions of the earth." In the seventeenth and eighteenth centuries, mathematician Philippe de La Hire studied instruments to measure climate, including temperature, pressure, and wind speed. He went on to collect data using these instruments at the Paris Observatory. In the nineteenth century, William Ferrel proposed a model for wind circulation, which was the first recorded theory to explain the westerly winds in the middle latitudes of both the northern and southern hemispheres. Ferrel cells are phenomena where air flows eastward and towards the pole near the Earth's surface, but westward and toward the equator at higher altitudes. The Beaufort wind scale was also named in the nineteenth century after Francis Beaufort, a British Rear Admiral who reportedly extended the work of many individuals in trying to standardize wind measurement and description. The invention of the cup anemometer by astronomer and physicist John Robinson in the middle of the same century aided in measuring winds and reputedly helped popularize the measure. The Beaufort wind scale was later revised by meteorologist George Simpson in the early twentieth century. Mathematician Lewis Richardson is widely considered a pioneer of mathematical weather prediction. He applied the method of finite differences and other mathematical methods in his Weather Prediction by Numerical Process in 1922. Wind is often mathematically modeled as a fluid, and some of Richardson's work was an extension of studies regarding water flow in peat. The Richardson number is a function involving gradients of temperature and wind velocity. Edward Milne, his contemporary, studied wind and sound, helping to refine huge binaural listening trumpets used to detected aircraft at night during World War I. In the twenty-first century, mathematicians often model various aspects of wind and wind power, including the wind movement through plant canopies using first and second order closure techniques; the probability of bird collisions with wind turbine rotors using statistical methods and calculus; descriptions and predictions of surface wind in mountainous terrain using statistical methods, geometry, vectors, and other mathematical functions; and the wind flow or turbulence over many types of surfaces, including turbine blades, ocean waves, automobiles, and structures.

U.S. Wind Research and Applications

The first wind system to generate electricity in the United States was built by Charles Brush in the late nineteenth century. However, there was relatively little development in that area until the energy crises of the 1970s, which motivated people to seek alternative sources of electricity, such as wind. The 1990s and the 2000s saw technological advances, decreasing turbine costs, and the emergence of popular and political support for wind energy. At the start of the twenty-first century, the U.S. government aimed to have 20% of all electricity generated by wind by 2030. Moreover, statistical studies and other data suggest that wind should be able to compete on a cost-effective basis with traditional fossil fuel sources. Some reports even estimate that wind will account for 26% of the increase in renewable energy production by 2035, though this extrapolation may not be reliable. Wind has shown a number of advantages compared to other forms of electricity production: it does not emit greenhouse gasses while in operation, it is freely available, it is not subject to energy security concerns, there are no waste products, and the maintenance costs are relatively low compared to traditional or nuclear generating facilities. For energy sources such as wind and nuclear, the emissions occur during the construction phase and tend to be associated with the amount of concrete and steel used in the facilities. Wind energy also faces technological problems with intermittency, as electricity can only be produced while the wind is blowing and this problem had been studied by mathematicians.

For example, the Weibull correlation model, based on the Weibull distribution named for mathematician (Ernst) Waloddi Weibull, estimates energy outputs with reduced uncertainty versus previous models, which is potentially useful for preventative operation and maintenance strategies. The National Renewable Energy Laboratory offers both wind data sets and has developed many mathematical models to explore wind energy grids, economic impact of wind energy, and even a model called Village Power Optimization Model for Renewables (ViPOR), which is a computational tool that facilitates the design of a village electrification system using the lowest cost combination of centralized and isolated power generation. Beyond land-based power generation, scientists and engineers like Maximillian Platzer and Nesrin Sarigul-Klijn are exploring the potential benefits of a return to wind energy as a supplement for large, ocean-going ships.

Wind Tunnels

Wind tunnels allow scientists and mathematicians to create wind under controlled conditions to test theories and applications. Mathematicians Benjamin Robins and George Cayley constructed simple spinning devices to model drag and other aerodynamic forces in the nineteenth century but the flow is difficult to control under such conditions. Engineer Francis Wenham is credited with the invention of the first enclosed wind tunnel, in 1871, with colleague John Browning. Wind tunnels were used by Orville and Wilbur Wright in developing their airplane prototypes as well as by German scientists at the famous World War II Peenemünde research facility. With advances in computer technology, the properties of wind are often modeled using computational fluid dynamics rather than physical data collection in wind tunnels, or the two methods are used to compare and cross-validate results. The foundations of these methods are the Navier–Stokes equations, which are systems of nonlinear partial differential equations developed by mathematicians Claude-Louis Navier and George Stokes.

Further Reading

Huler, Scott. *Defining the Wind: The Beaufort Scale and How a Nineteenth-Century Admiral Turned Science Into Poetry*. New York: Three Rivers Press, 2004.

Shepherd, William, and Li Zhang. *Electricity Generation Using Wind Power*. Singapore: World Scientific Publishing Company, 2010.

Walker, Gabrielle. *An Ocean of Air: Why the Wind Blows and Other Mysteries of the Atmosphere*. London: Bloomsbury, 2007.

Jason L. Churchill

Resource Guide

Books

Aaboe, Asger. *Episodes From the Early History of Mathematics*. Washington, DC: Mathematical Association of America, 1975.

Adrian, Yeo. *The Pleasures of Pi and Other Interesting Numbers*. Singapore: World Scientific Publishing, 2006.

Agresti, A. *Categorical Data Analysis*. Hoboken, NJ: Wiley, 2002.

Aho, A. V., J. E. Hopcrotf, and J. D. Ullman. *The Design and Analysis of Computer Algorithms*. Reading, MA: Addison-Wesley, 1976.

Albert, Jim, and Jay Bennett. *Curve Ball: Baseball, Statistics, and the Role of Chance in the Game*. New York: Springer-Verlag, 2001.

Ascher, Marcia. *Mathematics Is Everywhere: An Exploration of Ideas Across Cultures*. Princeton, NJ: Princeton University Press, 2002.

Ball, W. W. Rouse. *A Short Account of the History of Mathematics*. New York: Sterling Publishing Company, 2001.

Barnett, Raymond, Michael Ziegler, and Karl Byleen. *Calculus for Business, Economics, Life Science, and Social Science*. Upper Saddle River, NJ: Prentice-Hall, 2005.

Baumohl, Bernard. *The Secrets of Economic Indicators: Hidden Clues to Future Economic Trends and Investment Opportunities*. 2nd ed. Upper Saddle River, NJ: Pearson Education, 2008.

Beckmann, Petr. *A History of π (Pi)*. New York: Barnes & Noble, 1971.

Behrends, Ehrhard. *Five-Minute Mathematics*. Providence, RI: American Mathematical Society, 2008.

Bell, Eric Temple. *Men of Mathematics*. New York: Simon & Schuster, 1937.

Bennett, Jay, and James Cochran. *Anthology of Statistics in Sports*. Philadelphia, PA: Society for Industrial and Applied Mathematics, 2005.

Berggren, Lennart, Jon Borwein, and Peter Borwein. *Pi: A Source Book*. New York: Springer-Verlag, 1997.

Berlekamp, Elwyn R., John H. Conway, and Richard K. Guy. *Winning Ways for Your Mathematical Plays*. Natick, MA: AK Peters, 2001.

Blackwell, William. *Geometry in Architecture*. Hoboken, NJ: Wiley, 1984.

Blatner, David. *The Joy of π*. New York: Walker & Co., 1997.

Blue, Ron, and Jeremy White. *The New Master Your Money: A Step-by-Step Plan for Gaining and Enjoying Financial Freedom*. Chicago: Moody, 2004.

Blum, Raymond. *Mathemagic*. New York: Sterling Publishing, 1992.

Bodie, Zvi, Alex Kane, and Alan Marcus. *Investments*. Chicago, IL: McGraw-Hill/Irwin, 2008.

Borwein, Jonathan, and Peter Borwein. *A Dictionary of Real Numbers*. Pacific Grove, CA: Brooks/Cole Publishing Co., 1990.

Boyer, C. B. *A History of Mathematics*. Hoboken, NJ: Wiley, 1968.

Boyer, C. B. *The History of the Calculus and Its Conceptual Development*. New York: Dover Publications, 1949.

Brealey, Richard A., Stewart C. Myers, and Franklin Allen. *Principles of Corporate Finance*. 9th ed. New York: McGraw-Hill, 2008.

Bressoud, David. *The Queen of the Sciences: A History of Mathematics*. Chantilly, VA: The

Teaching Company, 2008.

Broverman, Samuel A. *Mathematics of Investment and Credit.* Winsted, CT: ACTEX Publications, 2008.

Burkett, Larry, and Brenda Armstrong. *Making Ends Meet: Budgeting Made Easy.* Gainesville, GA: Crown Financial Ministries, 2004.

Burton, David M. *The History of Mathematics: An Introduction.* New York: McGraw-Hill, 2005.

Calinger, Ronald. *A Contextual History of Mathematics.* Upper Saddle River, NJ: Prentice-Hall, 1999.

Clagett, Marshall. *Archimedes in the Middle Ages.* Madison: University of Wisconsin Press, 1964.

Closs, Michael. *A Survey of Mathematics Development in the New World.* Ottawa: University of Ottawa, 1977.

Closs, Michael, ed. *Native American Mathematics.* Austin: University of Texas Press, 1986.

Coe, Michael D. *Breaking the Maya Code.* New York: Thames and Hudson, 1992.

Copeland, Thomas E., J. Fred Weston, and Kuldeep Shastri. *Financial Theory and Corporate Policy.* 4th ed. Upper Saddle River, NJ: Pearson Education, 2005.

Cullen, Christopher. *Astronomy and Mathematics in Ancient China: The Zhou Bi Suan Jing.* Cambridge, England: Cambridge University Press, 1996.

Cuomo, Serafina. *Ancient Mathematics.* London: Routledge, 2001.

Davenport, Harold. *The Higher Arithmetic: An Introduction to the Theory of Numbers.* Cambridge, England: Cambridge University Press, 1999.

Davis, Morton D. *The Math of Money: Making Mathematical Sense of Your Personal Finances.* New York: Copernicus, 2001.

De Mestre, Neville. *The Mathematics of Projectiles in Sport.* Cambridge, England: Cambridge University Press, 1990.

Devlin, Keith. *The Math Gene: How Mathematical Thinking Evolved and Why Numbers Are Like Gossip.* New York: Basic Books, 2001.

———. *The Unfinished Game: Pascal, Fermat, and the Seventeenth-Century Letter That Made the World Modern.* New York: Basic Books, 2008.

Drobat, Stefan. *Real Numbers.* Upper Saddle River, NJ: Prentice-Hall, 1964.

Dudley, Underwood. *Numerology or What Pythagoras Wrought.* Washington, DC: Mathematical Association of America, 1997.

Eastway, Rob, and John Haigh. *Beating the Odds: The Hidden Mathematics of Sport.* London: Robson Books, 2007.

Eglash, Ron. *African Fractals: Modern Computing and Indigenous Design.* New Brunswick, NJ: Rutgers University Press, 1999.

Eves, Howard. *An Introduction to the History of Mathematics.* New York: Saunders College Publishing, 1990.

Flegg, G. *Numbers: Their History and Meaning.* New York: Schocken Books, 1983.

Friberg, Jöran. *Unexpected Links Between Egyptian and Babylonian Mathematics.* Singapore: World Scientific Publishing Co., 2005.

Friedman, Arthur. *World of Sports Statistics: How the Fans and Professionals Record, Compile, and Use Information.* New York: Athenaeum, 1978.

Fries, Christian. *Mathematical Finance: Theory, Modeling, Implementation.* Hoboken, NJ: Wiley, 2007.

Frumkin, Norman. *Guide to Economic Indicators.* Armonk, NY: M. E Sharpe, 2000.

Gamow, George. *One, Two, Three... Infinity.* New York: Viking Press, 1947.

Gardner, David, and Tom Gardner. *The Motley Fool Personal Finance Workbook: A Foolproof Guide to Organizing Your Cash and Building Wealth.* New York: Fireside Books, 2003.

Gardner, Martin. *Mathematics, Magic and Mystery.* New York: Dover, 1956.

Gay, Timothy. *The Physics of Football.* New York: HarperCollins, 2005.

Gerdes, Paulus. *Geometry From Africa: Mathematical and Educational Explorations.* Washington, DC: Mathematical Association of America, 1999.

Gillings, R. J. *Mathematics in the Time of the Pharaohs.* New York: Dover Publications, 1982.

Gutstein, Eric, and Bob Peterson, eds. *Rethinking Mathematics: Teaching Social Justice by the Numbers.* Milwaukee, WI: Rethinking Schools, 2005.

Hadamard, Jacques. *A Mathematician's Mind.* Princeton, NJ: Princeton University Press, 1996.

Hardy, G. H. *A Mathematician's Apology.* Cambridge, England: Cambridge University Press, 1941.

Henry, Granville C. *Logos: Mathematics and Christian Theology.* Lewisburg, PA: Bucknell University Press, 1976.

Hersh, Rueben. *What Is Mathematics, Really?* New York: Oxford University Press, 1997.

Hoyle, Joe Ben, Thomas F. Schaefer, and Timothy S. Doupnik. *Fundamentals of Advanced Accounting*. New York: McGraw-Hill, 2010.

Kalbfleisch, John D., and Ross L. Prentice. *The Statistical Analysis of Failure Time Data*. Hoboken, NJ: Wiley, 2002.

Katz, Victor J., ed. *Mathematics of Egypt, Mesopotamia, China, India, and Islam: A Sourcebook*. Princeton, NJ: Princeton University Press, 2007.

Kellison, Stephen G. *Theory of Interest*. New York: McGraw-Hill, 2009.

Kimmel, Paul D., Jerry J. Weygandt, and Donald E. Keiso. *Financial Accounting: Tools for Business Decision Making*. Hoboken, NJ: Wiley, 2009.

King, Jerry. *The Art of Mathematics*. New York: Plenum Press, 1992.

Klein, John P., and Melvin L. Moeschberger. *Survival Analysis: Techniques for Censored and Truncated Data*. New York: Springer-Verlag, 1997.

Kline, M., *Mathematical Thought From Ancient to Modern Times*. New York: Oxford University Press, 1972.

Koetsier, T., and L. Bergmans, eds. *Mathematics and the Divine: A Historical Study*. Amsterdam: Elsevier, 2005.

Longe, Bob. *The Magical Math Book*. New York: Sterling Publishing, 1997.

Martzloff, Jean-Claude. *A History of Chinese Mathematics*. New York: Springer-Verlag, 1987.

Moses, Robert P., and Charles E. Cobb, Jr. *Radical Equations: Civil Rights From Mississippi to the Algebra Project*. Boston: Beacon Press, 2001.

Mullis, Darrell, and Judith Handler Orloff. *The Accounting Game: Basic Accounting Fresh From the Lemonade Stand*. Naperville, IL: Sourcebooks, 2008.

Nahin, Paul J. *Dr. Euler's Fabulous Formula*. Princeton, NJ: Princeton University Press, 2006.

Nasar, Sylvia. *A Beautiful Mind: The Life of Mathematical Genius and Nobel Laureate John Nash*. New York: Simon & Schuster, 2001.

Oliver, Dean. *Basketball on Paper: Rules and Tools for Performance Analysis*. Washington, DC: Brassey's, 2004.

Pullan, J. M. *The History of the Abacus*. New York: F. A. Praeger, 1969.

Rafiquzzaman, M. *Fundamentals of Digital Logic and Microcomputer Design*. Hoboken, NJ: Wiley, 2005.

Rudin, W. *Principles of Mathematical Analysis*. New York: McGraw-Hill, 1953.

Salem, Lionel, Frédéric Testard, and Coralie Salem. *The Most Beautiful Mathematical Formulas*. Hoboken, NJ: Wiley, 1992.

Schwarz, Alan. *The Numbers Game: Baseball's Lifelong Fascination with Statistics*. New York: St. Martin's Press, 2004.

Smith, D. E. *History of Mathematics*. Vol. 2. New York: Dover Publications, 1958.

Solow, Daniel. *How to Read and Do Proofs: An Introduction to Mathematical Thought Process*. Hoboken, NJ: Wiley, 1982.

Steen, Lynn A. *On the Shoulders of Giants: New Approaches to Numeracy*. Washington, DC: National Academy Press, 1990.

Sterrett, Andrew. *101 Careers in Mathematics*. Washington, DC: The Mathematical Association of America, 1996.

Suzuki, Jeff. *A History of Mathematics*. Upper Saddle River, NJ: Prentice Hall, 2002.

Taylor, Alan D. *Mathematics and Politics: Strategy, Voting Power, and Proof*. New York: Springer-Verlag, 1995.

van der Waerden, B. L. *Geometry and Algebra in Ancient Civilizations*. Berlin: Springer, 1983.

Venema, G.A. *The Foundations of Geometry*. Upper Saddle River, NJ: Pearson Prentice Hall, 2006.

Weygandt, Jerry J., Paul D. Kimmel, and Donald E. Keiso. *Managerial Accounting: Tools for Business Decision Making*. Hoboken, NJ: Wiley, 2008.

Winkler, Peter. *Mathematical Puzzles: A Connoisseur's Collection*. Natick, MA: AK Peters, 2004.

Wright, Tommy, and Joyce Farmer. *A Bibliography of Selected Statistical Methods and Development Related to Census 2000*. Washington, DC: U.S. Bureau of the Census, 2000.

Yeldham, F. A. *The Teaching of Arithmetic Through Four Hundred Years (1535–1935)*. London: G. G. Harrap & Company, 1935.

Yong, L. L., and A. T. Se. *Fleeting Footsteps*. Singapore: Word Scientific Publications, 2004.

Zaslavsky, Claudia. *Africa Counts: Number and Pattern in African Culture*. Chicago: Lawrence Hill Books, 1999.

Zill, D. G. *Calculus with Analytic Geometry*. Boston: Prindle, Weber & Schmidt, 1985.

Journals and Magazines

The AMATYC Review

The American Mathematical Monthly

Association for Women in Mathematics Newsletter

Biometrics
Chance
The College Mathematics Journal
Experimental Mathematics
The Fibonacci Quarterly
Historia Mathematica
IMU-Net
Involve
Journal of Humanistic Mathematics
Journal of Integer Sequences
Journal of Recreational Mathematics
Journal of Statistics Education
Loci
MAA FOCUS
Math Horizons
Mathematics Magazine
Mathematics Teacher
NAM Newsletter
Notices of the American Mathematics Society
The Pentagon
Pi Mu Epsilon Journal
Plus Magazine
PRIMUS
Rose-Hulman Undergraduate Mathematics Journal
SIAM Review
Scholastic Math
Significance
Teaching Children Mathematics
Undergraduate Mathematics and Its Applications

Internet

American Institute of Mathematics
    www.aimath.org
The Algebra Project
    www.algebra.org
AMATYC
    www.amatyc.org
American Mathematical Society
    www.ams.org
American Statistical Association
    www.amstat.org
Association for Women in Mathematics
    www.awm-math.org
CryptoKids
    www.nsa.gov/kids
Datamath Calculator Museum
    www.datamath.org
Illuminations
    illuminations.nctm.org
MacTutor History of Mathematics
    www-history.mcs.st-and.ac.uk
Mathematical Fiction
    http://kasmana.people.cofc.edu/MATHFICT
Math for America
    www.mathforamerica.org
Math Forum
    www.mathforum.com
Math Fun Facts!
    www.math.hmc.edu/funfacts
MathDL
    mathdl.maa.org/mathDL
Mathematical Association of America
    www.maa.org
Mathematical Science Research Institute
    www.msri.org
The Museum of Mathematics
    www.momath.org
National Association of Mathematicians
    www.nam-math.org
National Council of Teachers of Mathematics
    www.nctm.org
RadicalMath
    www.radicalmath.org
Society for Industrial and Applied Mathematics
    www.siam.org
We Use Math
    www.weusemath.org
Wolfram MathWorld
    www.mathworld.wolfram.com

Index

Text and page numbers in **boldface** refer to main topics.

Abbott, Edwin
    Flatland, 77
acceleration, 81
Accelerator Mass Spectrometry, 22
accident reconstruction, 1–2
Achermann, Peter, 116
actuarial tables, 99
Adams, John Couch, 10, 49
Advanced Circulation Model (AD-CIRC), 87
aerodynamics, 4
Agassiz, Alexander, 35
Agreement Governing the Activities of States on the Moon and Other Celestial Bodies, 109
agricultural economics, 57
air resistance, 62, 63
Airy, George, 144
Akim, Efraim L., 109
albatross, 5
Aldrin, Edwin "Buzz", 94
All Around the Moon (Verne), 93
Almagest, The (Ptolemy), 117
American Society of Clinical Oncology, 27
Ampère, André-Marie, 126
Ampere's law, 126
Anastasi, Anne
    Psychological Testing, 124
Animalia kingdom, 2
animals, 2–8
    biological systematics, 2
    chimeras and hybrids, 7
    food webs, 6, 7
    migration and, 4, 5
    modeling based on, 3
    movement of, 4
    symmetry/fractals and, 7
    tissue structures, 3
Antikythera mechanism, 9
Apollo program, 94, 109
Arenstorf periodic orbits, 109
Arenstorf, Richard, 109
Aristotle
    Meteorologica, 156
Armstrong, Neil, 94
artillery
    types of, 61
Aspect, Alain, 101
assisted reproductive therapy (ART), 122
astrodynamics, 93
astronomy, 8–11
    astrophysics, 8
    during Renaissance, 9
    Greeks and, 9
    heliocentric, 137
    lunar calendars, 9
    parallax measurements in, 10
    satellites and, 132–163
autism, 19
automobiles
    radar guns and, 45

ballistics studies
    firearms and, 63
Barazangi, Muawia, 119
barometers, 49, 156, 159, 162
barometric pressure, 49
Barraquer, Jose, 152
basal body temperature, 59
bathyspheres, 87
Batty, David, 90
Bayesian decision theory, 122
Beaufort, Francis, 87, 160, 162
Beaufort wind scale, 87, 162
Becquerel, Alexandre-Edmond, 133
bees, 11–13
Bekenstein, Jacob D., 15
Bell inequalities, 101
Bell, John, 101
Benjamin-Lighthill conjecture, 145
Benjamin, Thomas, 145
Bentley, Wilson, 38
Berlekamp, Elwyn, 70
Bernoulli, Daniel
    mathematical modeling and, 88
    projectile trajectories and, 62, 63
Bernoulli, Johann (John, or Jean)
    projectile trajectories and, 62, 63
Berti, Gasparo, 159
Bessel, Friedrich, 10
Bhaumik, Mani Lal, 152
Big Bang theory, 45
Billings, Evelyn, 60
Billings, John, 60
Binet, Alfred, 90
bioinformatics, 72
biological systematics, 2
biomimicry, 138
biostatistics, 41
Birkoff, Garrett, 101

black holes, 13–15, 77, 127
Blake, John, 60
blood oxygen level dependence (BOLD), 18
blood pressure measurement, 136
body clocks, 116
Bohr, Niels, 101, 149
Boltzmann, Ludwig, 51, 143
Bolyai, János, 75
Bolyai-Lobachevskian geometry, 75
Bondi, Hermann, 53
bone movement, 95
Book of Signs, The (Theophrastus), 156
Brahe, Tycho, 10
brain, 15–20
    composition and structure, 16
    imaging technologies and, 18, 19
    nervous system and, 112
    nutrition and, 113
    plasticity, 92
brainbow, 18
Bravais, August, 37
Bravais lattice, 37
Brenner, Sydney, 16
Broca, Paul, 19
Broca's (frontal lobe), 19
Brush, Charles, 163
Bunsen, Robert, 10
Burning Index (BI), 67
business, economics, and marketing
    agricultural, 57
Byram, George, 67

Caenorhabditis elegans, 16
cancer, 27, 28
Caplan, Seth, 77
capsid geometry, 150
carbon dating, 20–22
carbon dioxide emissions, 33
carbon footprint, 22–25
carbon footprints
    calculation of, 22, 24
    economy/policy and, 24
    marginal abatement curve and, 24
    of people, 24
    per country, 25
careers
    emerging fields, 41
    medical and pharmaceutical, 41
cartography, 104
Cassini, Giovanni, 132

caves and caverns, 25–27
C. elegans, 16
celestial mechanics, 93
Celsius, Anders, 143
Celsius scale, 143, 159
Center for Bioinformatics, 27
center of gravity (CoG), 80
centigrade scale, 143
Chandrasekhar, Subrahmanyan (Chandra), 14, 126
chaos theories, 132
chemotaxis, 4
chemotherapy, 27–28
Chronbach, Lee, 125
classical test theory, 124
classic mathematical problems
    heat conduction problem, 142
Clausius, Rudolf, 52
Clean Water Act, 156
Clifford, William Kingdon, 52
climate change, 28–33
    as a distribution, 29
    global warming, 28, 30, 79
    sunspots, 136
climatology, 158
clouds, 34–35
cobweb theorem, 55
cognitive epidemiology, 92
Committee on Data for Science and Technology, 149
community-supported agriculture (CSA), 58
Conférence Général des Poids et Mesures, 100
Connes, Alain, 47
contact lenses, 152
contaminants, 154, 155
Conway, John H., 70
Copernicus, Nicolaus, 9
    De Revolutionibus, 118
    heliocentric astronomy and, 137
coral reefs, 35–37, 36
Coriolis, Gaspard-Gustave de, 145
Coriolis-Stokes force, 145
Cox, David, 146
Crick, Francis, 107
Crothers, Stephen, 13
crystallography, 37–38, 100

dancing
    bee choreography, 13

dark energy, 53
Darwin, Charles
    Structure and Distribution of Coral Reefs, 35
Deary, Ian, 90
deforestation, 38–41
Delambre, Jean, 132
Democritus of Abdera, 47
De Revolutionibus (Copernicus), 118
Descartes, René
    animal/machine connections and, 3
    mechanics and, 51
    tides/waves and, 144
    volcanos and, 153
diagnostic testing, 41–43
Dido of Tyre (Queen), 11
diets, 114
diffusion spectrum imaging, 18
diffusion tensor imaging (DTI), 18
digital elevation models, 49
diseases, tracking infectious, 92
disease survival rates, 43–44
DNA
    analysis, 70, 72
    electron microscopes and, 100
    molecular structure of, 106, 107
    viruses and, 150
domes, 154
Doppler, Christian, 45
doppler radar, 44–46
Dorman, James, 119
Dorsey, Noah, 100
dose-response curve, 96
Drexler, K. Eric, 110
drinking water, 154, 155
dry adiabatic lapse rate, 35
Dupain-Triel, Jean-Louis, 49
dynamical systems, 15
dyscalculia, 19

earthquakes, 46–47
eclipses, 8, 9, 108
ecosystems, 6
ecumene, 104
Eddington, Arthur, 14, 131
EEG/EKG, 18, 19
Ehlinger, Ladd, Jr., 77
Ehrenfest, Paul, 126
Einstein, Albert
    "E=MC2" album (Carey), 130
    gravity/weightlessness and, 160

light speed and, 100
photoelectric effect and, 134
quantum physics and, 139
Riemannian geometry and, 75
Theory of Everything, 78
theory of relativity and, 10, 13, 75, 129, 141
universal constants and, 148
electromagnetic wave equation, 100
elementary particles, 47–49
Elements of Physical Biology (Lotka), 121
elevation, 49–51
elliptical orbits, 74, 81, 118
Ellis, George, 13
energy, 51–53
Energy Conscious Scheduling (ECS), 82
Eniwetok Atoll, 35
entanglement, 101
epidemiology, 92
Eratosthenes of Cyrene, 9
Euclid of Alexandria
postulates of, 74
Eudoxus of Cnidus, 117
Eukaryota domain, 2
Eulerian-Lagrangian-Agent Method (ELAM), 155
Euler-Lotka equations, 59
event horizon, 13, 14
Exposition du Système du Monde (Laplace), 14
extinction, 53–55
eyeglasses, 151

Fahrenheit, Gabriel, 143
Fahrenheit temperature scale, 143, 159
Faraday, Michael, 126
Faraday's law of induction, 126
farming, 55–58, 56
Farr, William, 88
Feingold, Graham, 34
Ferrel, William, 162
fertility, 58–60
Feynman, Richard, 52, 101
Fields Medal, 49
fingerprints, 60–61
Finley, John, 86
firearms, 61–64
fireflies, 138
Fisher, Ronald, 55
fish schooling, 138
Fitzgerald, George, 100
Fizeau, Hippolyte, 45

Flatland (Abbott), 77
Flatland the Movie (movie), 77
flavonoids, 114
Fleischmann, Martin, 53
flight, animals, 2, 4, 5, 6
floods, 64–66
Fons, W. R., 66
food webs, 6
forest fires, 66–68
four-dimensional geometry, 100
Fourier, Jean Baptiste Joseph
heat conduction and, 142
On the Propagation of Heat in Solid Bodies, 142
Fourier series, 151
Fourier's heat equation, 142
Fourier transforms, 100
fractals
coral reefs and, 36
in animal kingdom, 7
lightning patterns and, 103
fraud detection
neural networks and, 16
Fraunhofer, Joseph von, 10
Frémont, John Charles, 49
Fresnel, Augustin-Jean, 100
Frisch, Karl von, 12
Fritts, Charles, 134
From the Earth to the Moon (Verne), 93
Frost, Wade Hampton, 89
Fujita-Pearson scale, 87
Fujita, Tetsuya Theodore, 86
Fuller, Buckminster "Bucky", 107
fullerenes, 107
functional MRI (fMRI), 18
fusion, 52

Galileo (Galileo Galilei)
animal/machine connections and, 3
invention of thermometer, 156
principle of relativity and, 129
telescopes and, 139
Galileo's principle of relativity, 129
Galton, Francis, 60
Game of Pistols, 70
game theory, 68–70
Gamow, George, 130
Gauss, Carl F.
curved space and, 131
laws for electricity and magnetism, 126

Gauss's laws for electricity and magnetism, 126
Gemini program, 94
genetically modified foods, 115
genetic engineering, 70
genetics, 70–73
genetic variability, 72
genotype, 71
GeoEye, 132
geometry of the universe, 73–78
dimensionality and, 77
Euclidean geometry, 77
global geometry and, 73
triangles and, 74
geosynchronous satellites, 133
geothermal electricity, 79
geothermal energy, 78–80
geothermal heating, 79
global geometry, 73, 76
globalization, 115
Goldman, David, 19
Goldman equation, 19
"Gompertzian growth" model, 27
GPS
satellites, 120, 131, 133
Granville, Evelyn Boyd, 109
graphs, 84, 113
Graunt, John, 98
gravitational time dilation, 131
gravity, 80–81, 149, 160
Greek mathematics
astronomy and, 9
greenhouse gases (GHGs), 79, 162
green mathematics, 82–84
Grossman, Sharon, 70
group theory, 8, 107
Grover, Lov, 101
growth charts, 84–85

Hales, Thomas, 11, 12
Halley, Edmund, 49, 98
Halley's comet, 162
Harriot, Thomas, 137
Hawking radiation, 15
Hawking, Stephen, 13, 15
heat conduction problem, 142
Heisenberg uncertainty principle, 149
Heisenberg, Werner, 149
hemoglobin, 106
Herschel, John, 162
Hertz (Hz), 148
Hess, Harry, 119

highly optimized tolerance (HOT), 103
Hipparchus of Rhodes, 9
Hodgkin, Alan, 19, 112
Hodgkin-Huxley equations, 112
Hohmann transfer orbit, 93
Hohmann, Walter, 93
Holmes, Arthur, 118
Hubble, Edwin, 45
Hubble Space Telescope, 45, 78, 132
Human Connectome Project, 17
Human Genome Project, 72
Hurricane Katrina, 86, 144
hurricanes and tornadoes, 85–88, 86
Huth, Andreas, 138
Huxley, Andrew, 19, 112
Huygens, Christiaan, 100, 116
Hyman, Albert, 116
Hyperbolic Crochet Coral Reef, 37
hyperbolic geometry, 37, 75

imaging technologies, 18
immunology, 114
Industrial Revolution
    catalysts, 162
    employment and, 56
    mass production and, 56
    steam engines and, 162
infectious diseases, tracking, 88–90
infinity
    universe and, 76
Institute for Figuring, 37
intelligence quotient (IQ)
    content of tests, 91
    correlative variables and, 91
    development of, 90
    link to health, 92
intelligence quotients (IQ), 90–92
Intergovernmental Panel on Climate Change (IPCC), 28
International Sun/Earth Explorer 3 (ISEE-3), 133
International System of Units (SI), 52
interplanetary travel, 92–94
item-response theory (IRT), 124

Jacques, Cassini, 132
James Webb Space Telescope, 78
Jeffrey, Harold, 46
joint loading, 96
joints, 95, 95–96
joule, 51
Joule, James, 51, 52

joule-seconds, 149

Kaku, Michio, 78
Kaplan, Edward, 146
Kelvin scale, 143
Kepler curve, 161
Kepler, Johannes
    laws of planetary motion, 10, 118, 133
Kepler's Laws, 10, 133
Kermack, W. O., 150
Kerr, Roy, 13
kinematic redundancy, 95
kinetic body data, 115
Kirchhoff, Gustav, 10
Kish, Leslie, 58
Kittinger, Joseph, 161
Klein, Felix
    Erlangen program, 75
Klemen, Michael, 162
Korotkoff, Nikolai, 136
Korotkoff sounds, 136
Kunz, Hanspeter, 117

Ladd, Harry, 35
Laennec, René, 135
Lagrange, Joseph Louis, 105
Lagrangian-Eulerian flow models, 78
La Hire, Philippe de, 162
Lambert, Johann Heinrich, 105
La nova scientia (Tartaglia), 62
Laplace, Pierre de
    Exposition du Système du Monde, 14
Larmor, Joseph, 100
laser-assisted in situ keratomileusis (LASIK), 152
lasers, 138, 139
LASIK, 152
lava domes, 154
laws of planetary motion, 10, 118, 133
LD50/median lethal dose, 96–98
Lebombo bone, 25
Leibniz, Gottfried Wilhelm
    energy and, 51
Leonardo da Vinci
    humidity measurements and, 156
Leslie, Nandi, 40
Le Système International d'Unités (SI), 52
lethality, 97
Leverrier, Urbain, 10

Libby, Willard, 20
Lichtman, Jeff, 18
Lidwell, Mark, 116
Lie groups, 48
Lie, Sophus, 48
life expectancy, 98–99
light, 99–101
light-emitting diodes (LEDs), 112
Lighthill, Michael, 132, 145
lightning, 101–103
Lind, James, 114
Linnæus, Carolus, 143
Lippershey, Hans, 139
Lorentz, Hendrik, 100
Lorentz transformations, 100, 130
Lotka, Alfred
    Elements of Physical Biology, 121
Lotka-Volterra system, 120

Maclaurin, Colin, 144
magnetic resonance imaging (MRI), 18
Magnetoencephalography (MEG), 19
magnetoperception, 4
malnutrition, 115
Mandelbrot, Benoit, 103
maps, 103–105
Marin, Mario, 127
Markov, Andrei, 147
Markov decision process model, 147
mathematical epidemiology, 122
mathematical modeling
    for accident reconstruction, 1
    for animals in nature, 3, 82, 120, 121
    data mining and, 40
    for dynamic systems, 36, 37, 40
    energy sustainability and, 82
    for firefly activity, 138
    flood predictions and, 65
    for forest fires, 66
    for geothermal processes, 78
    global warming and, 82
    for hurricanes and tornadoes, 86
    for infectious diseases, 89, 150
    for motion of moon, 108
    for nutrition, 114, 115
    for probability for survival, 97
    recycling and, 129
    SIR model for, 150
    for volcanos, 153, 154
    for water supplies, 155
    for wind, 157

Mathematicians of the African Diaspora, 40
Mathematics Awareness Month, 73
Maxwell-Boltzmann kinetic theory of gases, 143
Maxwell, James, 100, 126, 143
Mayer, J. C. A., 60
McKendrick, A.G., 150
McNamara, Robert, 70
meal planning, 115
medical imaging, 18, 95, 148
Meehl, Gerald, 30
Meier, Paul, 146
Mendel, Gregor, 71, 72
Mercury program, 94
Meteorologica (Aristotle), 156
meteorology, 45
Michell, John, 14
Michelson, Albert, 100
Michelson-Morley experiment, 130
microgravity, 161
microscopes, 100, 107
military code
    Morse code, 157
minerals, 114
minimax theorem, 68
Minkowski, Hermann, 100
Minkowski space, 100
Mirollo, Renato, 138
Miura, Koryo, 132
MODFLOW, 79
Modified Mercalli Intensity Scale, 46
molecular structure, 106–107
Monte Carlo simulation, 126
moon, 107–110
    eclipses, 8, 9, 108
    human exploration of, 109
    lunar calendars, 9
moonlanding, 109
Morse code, 157
Morse, Samuel F. B., 157
mortality as dose-response, 97
mortality tables, 99
Mouchot, Augustin, 134
Mount St. Helens, 153
Multi-Angle Imaging SpectroRadiometer, 34
Mutually Assured Destruction (MAD), 70

nanocars, 110
nanotechnology, 107, 110–112
nanotubes, 107

Napoleon Bonaparte (Napoleon III), 63
Nash Equilibria, 69
Nash, John, 69
National Aeronautics and Space Administration (NASA)
    Apollo program, 94, 109
    data collection/analysis, 29, 34
    elementary particles and, 48
    weightless research, 161
National Arbor Day Foundation, 30
National Cancer Institute, 27
National Center for Atmospheric Research (NCAR), 30
National Center for Electron Microscopy, 100
National Elevation Dataset, 50
National Fire-Danger Rating System (NFDRS), 67
National Institute of Biomedical Imaging and Bioengineering (NIBIB), 95
National Institutes of Health, 17
National Oceanic and Atmospheric Administration (NOAA), 29 144
National Renewable Energy Laboratory, 163
Native American mathematics
    weather forecasting and, 156
Navaratna, Channa, 116
Navaratna, Menaka, 116
Navier, Claude-Louis, 145
Navier-Stokes equations, 145
negative numbers
    topographical maps and, 50
Neptune, discovery of, 10
Nernst equation, 19
Nernst, Walther, 19
nervous system, 112–113
Neumann, John von, 142
neurochip, 113
NEURON computer simulation system, 112
neurons, 15, 16, 18
neuroscience, 113
Neutral Buoyancy Simulator, 161
New Principles of Gunnery (Robins), 63
Newtonian, 142
Newton, Sir Isaac
    animal/machine connections, 3
    laws of motion/laws of gravity and, 14, 118, 157
    light and, 99
    telescopes and, 142

non-Euclidean geometry, 10
Norse Greenland society, 32
Norton, Larry, 27
Norton-Simon hypothesis, 28
"number blindness" (dyscalculia), 19
numerical weather prediction, 157
nutrition, 113–116

Oceania, Pacific Islands, 119
Ohl, Russell, 134
On the Propagation of Heat in Solid Bodies (Fourier), 142
Operator Algebras, 101
Oppenheimer, J. Robert, 15
optics, 139
organ transplants, 146
orthonormal polynomials, 151
Outer Space Treaty, 109

pacemakers, 116–117
    body clocks/jet lag and, 116
    heart rhythms and, 116
Pacific Ring of Fire, 119
Packe, Christopher, 49
Pangea, 118
parabolic flight, 161
parallax measurements, 10
"Particle Zoo, The", 47
Pascal, Blaise
    barometric pressure/elevation and, 49
patterns
    caves and caverns, 25, 26
photovoltaic cells, 133
Pickover, Clifford, 91
Pierce, John, 132
Planck, Max, 101, 126, 149
planetary orbits, 117–118, 132, 133
plate tectonics, 118–120
Plato of Tivoli
    The Republic, 25
Platzer, Maximillian, 163
Poincaré, Jules Henri, 94, 100
polarized light, 100
Pons, Stanley, 53
population growth, 59
Posidonius of Rhodes, 162
positron-emission tomography (PET), 18
predator-prey models, 120–122
pregnancy, 122–123
prions, 106

Prisoner's Dilemma, 69
probability
    of survival, 97
Probability of Precipitation (PoP), 158
problem solving in society, 115
product-limit estimator, 146
psychological testing, 124–126
Psychological Testing (Anastasi), 124
psychometrics, 124
Ptolemy, Claudius
    Earth centered universe, 9, 74
    The Almagest, 117
Purkinje, Johannes, 60
Putin, Vladimir, 59
Pythagoras' cave, 25

quanta, 126
quantum computing, 101
quantum field theory (QFT), 101
quantum mechanics, 101, 106, 131
Questi et inventioni diverse (Tartaglia), 62

radiation, 126–127
"ramiform" pattern, 26
RAND Corporation, 69
Raven's Matrices, 91
reactive transport (RT) models, 155
recycling, 127–129
redshift, 45
Reed-Frost epidemic model, 89
Reed, Lowell, 89
refraction, 99
relativity, 129–131
Republic, The (Plato), 25
Resin Identification Codes, 128
resource intensity, 56
ribonucleic acid (RNA), 72
Richardson, Lewis Fry, 159, 162
Richter scale, 46
Riemann, Bernhard
    geometric formulations and, 75, 131
Riemannian geometry, 75
Ring of Fire, 78
RNA, 72
Robins, Benjamin, 62, 63
    New Principles of Gunnery, 63
robots, 94, 109
Röntgen, Wilhelm Conrad, 127
Rosa, Edward, 100
Ross, Ronald, 89
Rothermel, Richard C., 66

Royal Society of London, 14
Saffir, Herbert, 86
Saffir-Simpson scale, 86, 160
Sarigul-Klijn, Nesrin, 163
satellites, 132–133
    moon as, 107
scanning tunneling microscope, 107
Scheiner, Christoph, 137
Schmitt, Harrison H., 108
Schoenberg, Frederic, 68
Schor, Peter, 101
Schwabe, Heinrich, 137
Schwarzschild, Karl, 14, 130
scurvy, 114
seismic tomography, 26
Shipman, Barbara, 13
Simpson, George, 162
Simpson, Robert, 86
Single Photon Emission Computed Tomography (SPECT), 19
sinoatrial node (SA node), 116
SIR model, 150
SI units, 52
smallpox, 88
Smith, David, 60
Snell's Law, 100
snowflakes, 38
Snow, John, 88
solar panels, 133–135
Somayagi, Nilakantha, 49
sound telescopes, 141
spaceships, 93, 109
space-time geometry, 13
speleology, 25
spelunking, 25
sperm count, 60
sphygmometer, 136
Standard Model of Particle Physics, 47
Stanford-Binet test, 90
statistics
    survival analysis, 146
Steadman, Robert, 160
steam engines, 162
Stefan, Josef, 143
stethoscopes, 135–136
stochastic calculus, 79
Stokes, George, 145
stratus clouds, 34
Strogatz, Steven, 138
Structure and Distribution of Coral Reefs (Darwin), 35

sunspots, 136–137
superconducting quantum interference devices (SQUIDs), 19
survival analysis, 146
synchrony and spontaneous order, 137–139
Synthesis Report, 2007 (IPCC), 28
Système International d'Unités (SI), 52

Tartaglia, Niccolo
    La nova scientia, 62
    Questi et inventioni diverse, 62
taxonomies, 2
telegraphy, 157
telescopes, 139–142
temperature, 142–143
Temple, Blake, 53
tephra, 153
tessellations
    pattern blocks and, 11, 12
Thales of Miletus, 46
Theophrastus
    The Book of Signs, 156
Theory of Everything, 77, 78
theory of relativity, 10, 13, 129, 130, 141
thermodynamics, 15
thermometers, 143, 159
thermoscopes, 143
Thomson, William (Lord Kelvin), 143, 144
tides and waves, 144–146
tilings, 11, 12
time-to-pregnancy (TTP), 122
Tissot, Nicolas Auguste, 105
Tobacco Masaic Virus (TMV), 150
Toba volcano, 153
Topics in Mathematical Modeling (Tung), 31
topographic maps, 49, 50
Torricelli, Evangelista, 156, 159, 162
toxicity, 96
trajectories
    firearms and, 61, 62
transformations
    Lorentz, 130
transplantation, 146–147
Tung, Ka-Kit
    Topics in Mathematical Modeling, 31
Twersky, Victor, 126

ultrasound, 147–148

Union of Concerned Scientists, 132
United Nations Framework Convention on Climate Change, 38
United Nations World Population Prospects, 98
universal constants, 148–150
U.S. Army Signal Corps, 86
U.S. Centers for Disease Control and Prevention (CDC), 89
U.S. Department of Energy, 53
U.S. Geological Survey (USGS), 50
U.S. Joint Space Operations Center, 133
U.S. National Center for Health Statistics, 85
U.S. National Oceanic and Atmospheric Administration (NOAA), 34
U.S. Safe Drinking Water Act, 156

vaccination campaigns, 151
Van Allen belts, 94, 127, 132
Van Allen, James, 94, 127
Van de Hoop, Maarten, 118
van der Hilst, Robert, 118
velocity of light, 149
Verne, Jules
  All Around the Moon, 93
  From the Earth to the Moon, 93
Village Power Optimization Model for Renewables (ViPOR), 163
viruses, 150–151
vision correction, 151–153
vitamins, 114, 116
volcanos, 153–154
Volterra, Vito, 120
von Neumann, John, 68, 101

Walter Reed Army Medical Center, 146
War of the Worlds (Wells), 93
water quality, 154–156
waterwheels, 56
Watson, James D., 107
Watt, James, 162
wavelet analysis, 60
Wave Model (WAM), 87
wave theory of light, 100
weather forecasting, 86, 156–158
Weather Prediction by Numerical Process, 162
weather scales, 86, 158–160
Wechsler Adult Intelligence Scale, 90
Wechsler, David, 90
Wechsler Intelligence Scale for Children, 90
Wegener, Alfred, 118, 119
Weibull distribution, 163
Weibull, (Ernst) Waloddi, 163
weightless flight, 160–161
Weight Watchers Online program, 115
Wells, Herbert George (H.G.)
  War of the Worlds, 93
Wernicke, Carl, 19
Wernicke's (temporal lobe), 19

"white matter", 16
Wien, Wilhelm, 52, 126
Wilkins, J. Ernest, 53, 127
Wilkinson Microwave Anisotropy Probe (WMAP), 76
Wilson, Alexander, 137
wind and wind power, 162–163
wind tunnels, 163
wind turbines, 162
Wissel, Christian, 138
Witten, Edward, 49
Wolf, Rudolph, 137
World Health Organization (WHO), 88
wormholes, 15
Wu, Sijue, 145

X-rays, 106

Yellowstone caldera (supervolcano), 153
Yellowstone Volcano Observatory, 153
Yerkes Observatory, 141
yield, crop, 55
Young Choon Lee, 82
Young, Thomas, 100

Zero-G, 161
Zero Gravity Research Facility, 161
Zhang Heng, 46
Zomaya, Albert, 82